A First Course in Ordinary Differential Equations

A First Course in Ordinary Differential Equations

Fifth Edition

Walter Leighton

University of Missouri, Columbia

Wadsworth Publishing Company
Belmont, California
A division of Wadsworth, Inc.

821594

Mathematics editor: Richard Jones

Editing, design, production supervision: Brian Williams

Cover designer: Michael Rogondino

Copy editor: Winn Kalmon

Technical illustrator: John Foster

Printed in the United States of America

1 2 3 4 5 6 7 8 9 10—85 84 83 82 81

Library of Congress Cataloging in Publication Data

Leighton, Walter, 1907–
    A first course in ordinary differential equations.

    Bibliography: p.
    Includes index.
    1. Differential equations. I. Title.
QA372.L46   1980   515.3'52   80-23298
ISBN 0-534-00837-2

Contents

Preface to the Fifth Edition

This edition represents a major revision of its predecessor. A considerable amount of new material has been added, and in response to the suggestions of some who used the earlier book, the number of exercises has been increased by more than a third.

Special attention should, perhaps, be directed to Chapter 11, which now constitutes a rather complete introduction to Liapunov theory for autonomous systems that culminates in the application of the classical theorems of LaSalle and of Četaev. This material is supported by a large number of exercises that are within the capabilities of the student.

The author has always believed that textbooks should be written to be read by students, and a conscious effort has been made to present the material in such a way as to make this possible—without in any sense "writing down" to the student. It is hoped that teacher and student alike will find that this book affords a pleasant as well as a useful introduction to the area of ordinary differential equations.

W.L.

Preface to the Fourth Edition

This book has been written for the student who has completed a typical sequence of elementary calculus courses. I am aware that the word "typical" is not well-defined in this context, but perhaps it is sufficiently descriptive. Considerable pains have been taken to make this book one the student *can* understand—my goal throughout has been to write *for* the student.

A course in differential equations at this level is characteristically, and, I believe, desirably topical, and a variety of one-semester courses based on this book are possible. There is not a universal optimal ordering even of a chosen set of topics for such a course, but a textbook needs a unifying concept. In this book, the fundamental existence theorem—stated without proof in a simplified but adequately general form—serves to tie the various topics together. I believe that no understanding of even the most elementary theory can be acquired without some knowledge and experience with this central theorem.

Proofs are provided when their understanding is believed to be within the capabilities of the student at this level. The more sophisticated proofs are omitted.

Most of the applications appear in Chapter 5. The instructor will undoubtedly wish to vary his attention to this chapter according to the needs and interests of his class of the moment. It may be heresy in these times to say so, but this chapter can be omitted. I myself would find it difficult to omit the part of Chapter 5 that is based on Newton's laws, however disinterested a given class might happen to be in the applications. Newton's laws are not only basic in mechanics, but they surely must be regarded as an important segment in our liberal arts heritage.

A number of formulas are provided in Chapter 8 for the approximate integration of first-order differential equations. Section 4 of that chapter can be omitted. New useful formulas are given in Section 4 for the approximation of a conjugate point. Section 4 of Chapter 9 provides a summary of very elementary operations with matrices for students who have not had a course in linear algebra.

Chapter 10 contains the usual material on the phase plane for linear systems with constant coefficients; however, the treatment of the case when the characteristic roots are conjugate complex numbers is new. The elements of Liapunov theory appear in Chapter 11. A minimal amount of oscillation theory, including the Sturm separation and comparison theorems, is given in Chapter 6.

Exercises marked with asterisks are more difficult than the others. These appear throughout the book.

My debt to many students and colleagues, past and present, is very large indeed. Special thanks are due to Paul Burcham, Marc Q. Jacobs, C. B. Lynch, C. A. Swanson, David V. V. Wend, and Paul Williams for their many helpful suggestions as the manuscript for the present book was being prepared.

Walter Leighton

A First Course in Ordinary Differential Equations

1

Elementary methods

1 Introduction

Differential equations are equations that involve derivatives. For example, the equations

$$y' = f(x),$$

$$y'' + y = 0,$$

(1.1)
$$y'' = (1 + y'^2)^{1/2},$$

$$\frac{\partial^2 u}{\partial x^2} + \frac{\partial^2 u}{\partial y^2} = 0$$

are differential equations. The first three of these equations are called *ordinary* differential equations because they involve the ordinary derivatives of the unknown y. The last equation is an example of a partial differential equation.

We shall be concerned with ordinary differential equations and their solutions. Unless otherwise specified, we shall be dealing with differential equations having real coefficients, and we shall be discussing real solutions of such equations.

To solve an algebraic equation, such as

$$(1.2) \qquad x^2 - 3x + 2 = 0,$$

we seek a number with the property that when the unknown x is replaced by this number the left-hand member of the equation reduces to zero. In equation (1.2) either the number 1 or the number 2 has this property. We say that this equation has the two solutions 1 and 2. To solve a differential equation we seek to determine not an unknown number but an unknown function. For example, in the equation

$$(1.3) \qquad y'' + y = 0,$$

y is regarded as the unknown. To find a *solution*† we attempt to determine a *function defined on an interval* with the property that, when y is replaced by this function, the equation reduces to an identity on this interval. It is clear that $\sin x$ is a solution of (1.3) for all values of x, for

$$(\sin x)'' + \sin x \equiv 0 \qquad (-\infty < x < \infty).$$

Similarly, it is easy to verify that $\cos x$ is also a solution of the differential equation (1.3).

Differential equations play a fundamental role in almost every branch of science and engineering. They are of central importance in mathematical analysis. A differential equation describes the flow of current in a conductor; another describes the flow of heat in a slab. Other differential equations describe the motion of an intercontinental missile; still another describes the behavior of a chemical mixture. Sometimes it is important to find a particular solution of a given differential equation. Often we are more interested in the existence and qualitative behavior of solutions of a given differential equation than we are in finding its solutions explicitly.

In this chapter we shall begin our study by solving certain simple and important types of differential equations.

The *order* of a differential equation is the order of the highest derivative that appears in the equation. Accordingly, the first equation in (1.1) is of first order, and the next two equations are of second order. Similarly, the differential equation

$$y''' + y^4 = e^x$$

is of third order, and the equation

$$(y'''')^2 + yy' = 3$$

---

† We shall follow the custom of *italicizing* words and phrases that are being defined either explicitly or implicitly in the text.

is of fourth order. The differential equation

(1.4) $$M(x, y) + N(x, y)y' = 0$$

is of first order. It is frequently useful to rewrite this equation in the form

(1.4)' $$M(x, y)\, dx + N(x, y)\, dy = 0.$$

Thus,

$$(x^2 + y^2)\, dx + 2x\, dy = 0,$$

$$xe^y\, dx + (1 + y)\, dy = 0$$

are differential equations of first order written in the form (1.4)'.

Exercises

1. Verify that if c_1 and c_2 are constants, $c_1 \sin x + c_2 \cos x$ is a solution of the differential equation $y'' + y = 0$.

2. Find by inspection a solution of each of the following differential equations:
   (a) $y' - y = 0$;
   (b) $y' + 2y = 0$;
   (c) $y' = \sin x$;
   (d) $y' = e^{2x}$.

3. Find by inspection a solution of each of the following differential equations:
   (a) $y'' - y = 0$;
   (b) $xy' - y = 0$;
   (c) $y'' = 0$;
   (d) $y'' = 6x$.

4. Verify that the function $c_1 e^x + c_2 e^{2x}$ (c_1, c_2 constants) is a solution of the differential equation $y'' - 3y' + 2y = 0$.

5. Verify that $c_1 x + c_2 x^2$ (c_1, c_2 constants) is a solution of the differential equation $x^2 y'' - 2xy' + 2y = 0$.

6. Determine $p(x)$ so that the function $\sin \ln x$ ($x > 0$) is a solution of the differential equation $(xy')' + p(x)y = 0$.

7. Verify that $\sin x$ is a solution of the differential equation $y'^2 + y^2 = 1$.

8. Verify that if c_1 and c_2 are constants and $x > 0$, the function

$$c_1 \sin \frac{1}{x} + c_2 \cos \frac{1}{x}$$

   is a solution of the differential equation $(x^2 y')' + x^{-2} y = 0$.

9. Verify that $\sin x$, $\cos x$, $\sinh x = \frac{1}{2}(e^x - e^{-x})$, and $\cosh x = \frac{1}{2}(e^x + e^{-x})$ are solutions of the differential equation $y'''' - y = 0$.

Answers

2. (a) e^x.

6. $\dfrac{1}{x}$.

2 Linear differential equations of first order

A linear differential equation of first order is an equation that can be put in the form

(2.1) $$k(x)y' + m(x)y = s(x).$$

On intervals on which $k(x) \neq 0$, both members of this equation may be divided by $k(x)$, and the resulting equation has the form

(2.2) $$y' + a(x)y = f(x).$$

We shall suppose that $a(x)$ and $f(x)$ are continuous on some interval I.

Throughout this book an *interval I* will mean one of the following:

(2.3)
$$
\begin{aligned}
&[a, b] = a \le x \le b \quad \textit{closed interval,} \qquad &&(-\infty, b] = -\infty < x \le b, \\
&(a, b) = a < x < b \quad \textit{open interval,} \qquad &&(-\infty, b) = -\infty < x < b, \\
&[a, b) = a \le x < b, \qquad &&[a, \infty) = a \le x < \infty, \\
&(a, b] = a < x \le b, \qquad &&(a, \infty) = a < x < \infty, \\
& \qquad &&(-\infty, \infty) = -\infty < x < \infty.
\end{aligned}
$$

In (2.3), a and b are, of course, finite numbers.

When studying the differential equation (2.2) it is frequently useful to consider its *associated homogeneous* equation

$$y' + a(x)y = 0.$$

We have the following useful result.

Theorem 2.1. *If $a(x)$ is continuous on an interval I, the general solution of the first-order, linear, homogeneous differential equation*

(2.4) $$y' + a(x)y = 0$$

is†

(2.5) $$ce^{-\int a(x)\,dx} \qquad \textit{(c constant)}.$$

Further, this solution is valid on the entire interval I.

---

† In this book, unless otherwise indicated, we shall employ the symbol
$$\int f(x)\,dx$$
to mean any particular function $g(x)$ whose derivative is $f(x)$.

To illustrate the theorem, consider the differential equation

(2.6) $(1 + x^2)y' + 2xy = 0$ $(-\infty < x < \infty)$.

First, we divide through by the quantity $(1 + x^2)$ to put the equation in the form of (2.4):

$$y' + \frac{2x}{1 + x^2}\, y = 0.$$

Then, $a(x) = 2x/(1 + x^2)$, and (2.5) yields the solution

$$ce^{-\int \frac{2x\, dx}{1+x^2}} = ce^{-\ln(1 + x^2)} = \frac{c}{1 + x^2}   (c \text{ constant}).$$

That this function is a solution of the given differential equation for each choice of the constant c may be readily verified by substitution in the differential equation. It will be just as easy, however, to substitute (2.5) into (2.4) and begin our proof of the theorem. To this end, note first that

(2.7) $$\frac{d}{dx}\, e^{-\int a(x)\, dx} = -a(x)e^{-\int a(x)\, dx};$$

accordingly, if we set

$$y = ce^{-\int a(x)\, dx}   (c \text{ constant})$$

we have at once that

$$y' + a(x)y \equiv 0   (\text{on } I).$$

We have proved half of the theorem—namely, that (2.5) is a solution of (2.4) on I for every choice of the constant c. To be able to say that (2.5) is the *general solution* we must prove a converse: that every solution of (2.4) on I may be put in the form (2.5). So suppose that the function $y(x)$ is a solution of (2.4). Then,

$$y'(x) + a(x)y(x) \equiv 0   (\text{on } I).$$

We may multiply both members of this identity by the function

$$e^{\int a(x)\, dx}$$

obtaining

$$e^{\int a(x)\, dx}y'(x) + e^{\int a(x)\, dx}a(x)y(x) \equiv 0,$$

or

$$[e^{\int a(x)\, dx}y(x)]' \equiv 0.$$

It follows that if $y(x)$ is a solution,

$$e^{\int a(x)\, dx}y(x) \equiv c,$$

where c is a constant.

The proof is complete, for then

$$y(x) \equiv ce^{-\int a(x)\,dx}.$$

There are two commonly used elementary methods for solving an equation of the form (2.2). We begin by illustrating one of them.

Method 1.   The differential equation

(2.8) $$y' + y = 1 + e^{-x}$$

is of the form (2.2). Here, $a(x) = 1$, $f(x) = 1 + e^{-x}$. If we multiply both members of this equation by the function

$$e^{\int a(x)\,dx} = e^x,$$

the left-hand member becomes

$$(e^x y)',$$

and the equation (2.8) becomes

$$(e^x y)' = e^x + 1.$$

It follows that a solution y is given by

$$e^x y = \int (e^x + 1)\,dx = e^x + x + c \qquad (c\ \text{constant}),$$

or

$$y = 1 + xe^{-x} + ce^{-x}.$$

In this example, the interval I may be any interval on the x-axis—or, indeed, all of it. Note that $1 + xe^{-x}$ is a particular solution of (2.8) while ce^{-x} is the general solution of the associated homogeneous equation.

Here is a slightly more complicated example.

Example.   Solve the differential equation

(2.1)′ $$x^2 y' + xy = 2 + x^2 \qquad (x > 0).$$

We first put this equation in the form (2.2) by dividing it through by x^2:

(2.2)′ $$y' + \frac{1}{x} y = \frac{2}{x^2} + 1.$$

Here $a(x) = 1/x$ and $f(x) = 1 + 2/x^2$. We note that

$$e^{\int a(x)\,dx} = e^{\int \frac{dx}{x}} = e^{\ln x} = x.$$

If both members of (2.2)′ are multiplied by x, we have

(2.9) $$(xy)' = \frac{2}{x} + x.$$

From (2.9) we have successively

$$xy = 2 \ln x + \frac{x^2}{2} + c,$$

and

(2.10) $$y = \frac{2}{x} \ln x + \frac{x}{2} + \frac{c}{x} \qquad (x > 0).$$

It is easy to verify that this method of solving a linear differential equation of first order is always available, for if we multiply both members of equation (2.2) by the *integrating factor*

$$e^{\int a(x)\,dx},$$

that equation becomes

(2.11) $$[e^{\int a(x)\,dx}y]' = f(x)e^{\int a(x)\,dx}.$$

To solve (2.11) for y, we write

$$e^{\int a(x)\,dx}y = c + \int f(x)e^{\int a(x)\,dx}\,dx \qquad (c \text{ constant}),$$

and, finally,

(2.12) $$y = e^{-\int a(x)\,dx}\left[c + \int f(x)e^{\int a(x)\,dx}\,dx\right].$$

The student is advised to use the method described for solving an equation (2.2) rather than formula (2.12).

The derivation of (2.12) indicates that if there is a solution of (2.2), it can be put in the form of (2.12). Conversely, the substitution of (2.12) in (2.2) demonstrates that the right-hand member of (2.12) is indeed a solution of this differential equation over the (entire) interval I. We have then the following theorem.

Theorem 2.2. *If $a(x)$ and $f(x)$ are continuous on an interval I, the general solution of the differential equation*

$$y' + a(x)y = f(x)$$

is given by

$$e^{-\int a(x)\,dx}\left[c + \int f(x)e^{\int a(x)\,dx}\,dx\right] \qquad (c \text{ constant}),$$

and this solution is valid on the entire interval I.

Exercises

Solve each of the following differential equations:

1. $y' + y = 3$.

2. $y' - 2y = x$.

3. $xy' + y = \cos x + e^x$.

4. $x^2 y' - xy = x^4 + 3$.

5. $y' + ay = b$ $(a, b \text{ constants})$.

6. $y' + \dfrac{1}{x} y = e^{x^2}$ $(x > 0)$.

7. $y^2\, dx + (y^2 x + 2xy - 1)\, dy = 0$. (*Hint.* Reverse the roles of x and y.)

8. $y + (x - y^3 - 2)y' = 0$.

In the next four exercises find solutions $y(x)$ of the given differential equations that satisfy the given conditions.

9. $y' - y = 1,\ y(0) = 0$.

10. $xy' + y = 3x^2,\ y(1) = 1$.

11. $e^{-x}y' + 2e^x y = e^x,\ y(0) = \dfrac{1}{2} + \dfrac{1}{e}$.

12. $(\sin x)y' + (\cos x)y = \cos 2x,\ y\left(\dfrac{\pi}{2}\right) = \dfrac{1}{2}$.

Answers

1. $y = 1 + ce^{-x}$.

3. $\dfrac{\sin x}{x} + \dfrac{1}{x} e^x + \dfrac{c}{x}$.

5. $y = \dfrac{b}{a} + ce^{-ax}$.

7. $x = y^{-2}[1 + ce^{-y}]$.

9. $e^x - 1$.

11. $e^{-e^{2x}} + \frac{1}{2}$.

Method 2. Before proceeding with a second method for solving linear, first-order, differential equations it will be helpful to have available the following corollary to Theorem 2.1.

Corollary.   If $y_0(x)$ is any particular solution of the differential equation

(2.2) $$y' + a(x)y = f(x),$$

its general solution is

(2.13) $$y_0(x) + ce^{-\int a(x)\,dx}   \quad (c \text{ constant}).$$

That is, the general solution of (2.2) is the sum of any particular solution of that equation and the general solution of its associated homogeneous equation. For example, the differential equation

(2.14) $$y' - y = -3$$

has, by inspection, as a particular solution the function 3. The associated (or *corresponding*) homogeneous equation

$$y' - y = 0$$

has ce^x as its general solution. The corollary then asserts that the general solution of (2.14) is

$$3 + ce^x.$$

To prove the corollary, note first that the substitution of (2.13) in equation (2.2) shows that (2.13) is indeed a solution of (2.2) for every choice of the constant c.

To prove the converse, suppose $y(x)$ is an arbitrary solution of (2.2). We have then the two identities

$$y'(x) + a(x)y(x) \equiv f(x),$$
$$y_0'(x) + a(x)y_0(x) \equiv f(x).$$

Subtracting the second identity from the first yields

$$[y(x) - y_0(x)]' + a(x)[y(x) - y_0(x)] \equiv 0;$$

that is, the difference $y(x) - y_0(x)$ is a solution of the associated homogeneous equation (2.4). It follows from Theorem 2.1 that

(2.7) $$y(x) - y_0(x) \equiv ce^{-\int a(x)\,dx},$$

and the proof of the corollary is complete.

Let us return to the equation

(2.8) $$y' + y = 1 + e^{-x},$$

and consider its associated homogeneous equation

$$y' + y = 0.$$

One solution of the latter is clearly e^{-x}. We now introduce a new variable v by substituting

(2.15)
$$y = e^{-x}v$$

in (2.8). We have, then, that

$$(e^{-x}v)' + e^{-x}v = 1 + e^{-x},$$

or

$$e^{-x}v' = 1 + e^{-x}.$$

Accordingly,

$$v' = e^x + 1,$$

and one choice of v is $e^x + x$. It follows from (2.15) that a particular solution of (2.8) is

$$1 + xe^{-x}.$$

Using the corollary to Theorem 2.1 we have, then, that the general solution of (2.8) is

$$1 + xe^{-x} + ce^{-x}.$$

It is not difficult to see that this method also always solves the differential equation

(2.12)
$$y' + a(x)y = f(x),$$

since the function

$$e^{-\int a(x)\,dx}$$

is a solution of the associated homogeneous equation, and the substitution in (2.12) of

$$y = e^{-\int a(x)\,dx}v, \qquad y' = e^{-\int a(x)\,dx}[v' - a(x)v]$$

yields

$$v' = f(x)e^{\int a(x)\,dx}.$$

From this last equation we have

$$v = c + \int f(x)e^{\int a(x)\,dx}\,dx,$$

and, hence, that

$$y = e^{-\int a(x)\,dx}\left[c + \int f(x)e^{\int a(x)\,dx}\,dx\right],$$

which agrees with (2.12).

We have been taking the independent variable to be x in the above discussion. Any letter may be used, of course. A popular and useful choice is also the letter t, particularly when it represents *time*. To solve, for example, the differential equation

$$(2.16) \qquad (1 + t)\frac{dx}{dt} + 2x = 3(1 + t) \qquad (t > -1)$$

for x as a function of t we note first that the equation is a first-order linear differential equation in x and its derivative. We next divide through by the quantity $(1 + t)$ and write

$$(2.17) \qquad \frac{dx}{dt} + \frac{2}{1 + t} x = 3.$$

An integrating factor is then

$$e^{\int \frac{2dt}{1+t}} = e^{2\ln(1+t)} = (1 + t)^2.$$

Equation (2.17) may then be written as

$$(2.18) \qquad \frac{d}{dt}[(1 + t)^2 x] = 3(1 + t)^2,$$

and integrating we have

$$(1 + t)^2 x = (1 + t)^3 + c \qquad (c \text{ constant}),$$

or

$$x = 1 + t + \frac{c}{(1 + t)^2} \qquad (t > -1).$$

When t is the independent variable it is customary to write $\dot{x} = dx/dt$. Thus, equation (2.18), for example, might equally well have been written

$$[(1 + t^2)x]^{\boldsymbol{\cdot}} = 3(1 + t)^2.$$

Exercises

Solve the following differential equations by Method 2.

1. $y' + 2y = 4.$

2. $y' + y = x.$

3. $xy' + y = -\sin x - e^x.$

4. $x^2 y' - xy = x^3 + 4.$

Solve the following differential equations by Method 1.

5. $\dot{x} + \frac{1}{t} x = 4 \cos t^2.$

6. $\dot{x} + \dfrac{1}{1 + 2t}\, x = 4.$

7. $\dfrac{dx}{dt} + \dfrac{1}{125 + 3t}\, x = 12.$

Solve the following differential equations by either method.

8. $xy' - 3y = 4x^4 \sin 2x.$

9. $xy' + 2y = 3 \cos x^2.$

10. $\dot{x} + 2x = 4.$

11. $\dot{x} - 3x = 2e^{2t}.$

12. $(\tan x)y' + y = -3xe^x \sec x.$

13. $(\sin x)y' \times (\sec x)y = \cos^2 x \csc x.$ (*Hint.* It will be helpful to note that $\sec x \csc x = \sec^2 x/\tan x$.)

14. $\dot{x} - 2x = 2te^{t^2 + 2t}.$

15. $y' + y = 2,\ y(0) = 0.$

16. $t\dot{x} + x = \cos t,\ x(\pi) = 1.$

17. $\dot{x} + \dfrac{1}{125 + 3t}\, x = 12,\ x(0) = 2{,}000$ (see Exercise 7).

18. $xy' - y = x^2 + 3,\ y(1) = 0.$

19. $y' + \dfrac{\cos x}{\sin x}\, y = 2 \cos x,\ y\!\left(\dfrac{\pi}{2}\right) = 0.$

20. $y' + 3y = 4e^x,\ y(0) = 2.$

21. $xy' - 2y = x^3 e^{2x},\ y(1) = \tfrac{1}{2}.$

22. $y' + (\cot x)y = e^{2x} \csc x,\ y\!\left(\dfrac{\pi}{2}\right) = \tfrac{3}{2}\, e^{\pi}.$

23. $(x \ln x)y' + y = 2x^2,\ y(e) = 1 + e^2.$

24.* Show that the substitution

$$u = y^{1-n}$$

reduces the so-called *Bernoulli* equation

(1) $\qquad\qquad y' + p(x)y = q(x)y^n \qquad (n \neq 1)$

to the linear equation

$$\frac{1}{1-n}\, u' + p(x)u = q(x).$$

(*Hint.* First multiply both members of equation (1) by y^{-n}.)

25. Solve the Bernoulli equation

$$y' - y = xy^{1/2}.$$

26. Solve the Bernoulli equation

$$\frac{dx}{dy} + x = x^2.$$

27. Solve the Bernoulli equation

$$\frac{dx}{dt} - x = x^{3/2} \sin t$$

28. Solve the Bernoulli equation

$$y' + 2y = xy^2.$$

29. Solve

$$\dot{x} + x = -\tfrac{2}{3} e^{2t}x^4, \qquad x(0) = -1.$$

30. Solve

$$y' + (\cot x)y = -(\csc x)y^2, \qquad y\left(\frac{\pi}{4}\right) = \frac{1}{\sqrt{2}}.$$

31. Determine $r(x)$ so that $\sin x^2$ is a solution of the differential equation

$$[r(x)y']' + 2xy = 0.$$

Answers

1. $2 + ce^{-2x}$.

3. $(\cos x - e^x + c)/x$.

5. $(2 \sin t^2 + c)/t$.

7. $3(125 + 3t)^{4/3} + c(125 + 3t)^{-1/3}$.

9. $\tfrac{1}{2}x^{-2} (3 \sin x^2 + c)$.

11. $-2e^{2t} + ce^{3t}$.

13. $\cot x [\ln|\sin x| + c]$.

15. $2 - 2e^{-x}$.

16. $(\pi + \sin t)/t$.

17. $3(125 + 3t)^{4/3} + 625(125 + 3t)^{-1/3}$.

18. $x^2 + 2x - 3$.

19. $-\dfrac{\cos^2 x}{\sin x}$.

21. $\frac{1}{2}x^2(e^{2x} + 1 - e^2)$.

23. $\dfrac{1 + x^2}{\ln x}$.

25. $(ce^{x/2} - x - 2)^2$.

27. $(ce^{-t/2} - \frac{1}{5}\sin t + \frac{2}{5}\cos t)^{-2}$.

29. $(e^{3t} - 2e^{2t})^{-1/3}$.

3   Some applications

In this section we consider some typical applications of linear differential equations of first order. The equations we shall consider are of the form

$$\frac{dx}{dt} = b + f(t)x,$$

where b is a constant, possibly zero, and $f(t)$ is either always positive or always negative. In both these cases the change in x is roughly proportional to the amount x present at time t [when $b = 0$ and $f(t)$ is constant, it will be observed from the differential equation that the change in x is precisely proportional to the amount x present at time t].

Mixtures.   We are now in a position to solve a typical problem concerning mixtures. A large tank contains 100 gal. of brine in which 10 lb. of salt is dissolved. Brine containing 0.2 lb. of salt to the gallon runs into the tank at the rate of 6 gal./min. The mixture, kept uniform by stirring, runs out of the tank at the rate of 4 gal./min. Find the amount of salt in solution in the tank at the end of t min.

Let x be the number of pounds of salt in solution after t min., and note that the tank will contain $100 + (6 - 4)t$ gal. of brine at this time. The concentration will then be

$$\frac{x}{100 + 2t} \text{ lb./gal.}$$

Thus,

(3.1) $$\frac{dx}{dt} = (0.2)(6) - \frac{x}{100 + 2t}\, 0.4 \text{ lb./min.};$$

that is, the rate at which the amount of salt in solution is changing at time t is the amount per minute entering into solution minus the rate at which the salt leaves the tank. Equation (3.1) is linear, and we seek the solution with the property that $x = 10$ when $t = 0$. This solution may be obtained using the methods of Section 2. It is seen to be

$$x = 0.4(50 + t) - \frac{25(10^3)}{(50 + t)^2} \text{ lb.}$$

After 50 min., for example, there will be 37.5 lb. of salt in solution in the tank.

On units. Suppose the tank originally held 100 l. of brine in which 10 kg. of salt were dissolved, and that brine containing 0.2 kg. of salt to the liter ran into the tank at the rate of 6 l. per minute. Suppose further that the mixture, kept uniform by stirring, runs out of the tank at the rate of 4 l. per minute.

If x is the number of kilograms of salt in solution after t minutes, we have that the concentration at time t will be

$$\frac{x}{100 + 2t} \text{ kg./l.,}$$

and that

(3.1)′ $$\frac{dx}{dt} = 0.2(6) - \frac{x}{100 + 2t} 4 \text{ kg./min.}$$

Again, the solution of this differential equation will be

$$x = 0.4(50 + t) - \frac{25(10^3)}{(50 + t)^2} \text{ kg.,}$$

and after 50 min., for example, there will be 37.5 kg. of salt in the tank.

This pair of examples illustrates that although units may often be introduced usefully into mathematical computation, mathematical analysis itself is concerned only with pure numbers. If in the last example, liters were replaced by jeroboams, kilograms by dozens, salt by insects, minutes by days, after 50 days there would be 37.5 dozens of insects in the tank.

Decompositions. Suppose that a quantity of a radioactive substance originally weighing x_0 grams decomposes at a rate proportional to the amount present and that half the original quantity is left after a years. To find the amount x of the substance remaining after t years, note that we may solve the differential equation

(3.2) $$\frac{dx}{dt} = -kx \qquad (k \text{ constant}),$$

subject to the condition that $x = x_0$ when $t = 0$. This equation is linear, and the solution which satisfies the given condition is

(3.3) $$x = x_0 e^{-kt}.$$

The constant of proportionality k is then determined by the equation

$$\frac{x_0}{2} = x_0 e^{-ak},$$

or $e^{-k} = (\frac{1}{2})^{1/a}$. Thus, equation (3.3) becomes

(3.4) $x = x_0 (e^{-k})^t = x_0 (\frac{1}{2})^{t/a},$

and we note that the number x of grams of the substance remaining un-decomposed decreases exponentially.

Equation (3.4) can now be employed to determine, for example, the number of years that would be required for the substance to be reduced to one-tenth the original amount. Suppose half the substance remains after, say, 1,000 years. We then have to solve the equation

$$\tfrac{1}{10} x_0 = x_0 (\tfrac{1}{2})^{t/1,000},$$

or

$$(\tfrac{1}{2})^{t/1,000} = \tfrac{1}{10}.$$

Taking logarithms of both members of this equation leads to

$$t = \frac{1,000}{\log 2} = 3,300 \text{ yr. (approx.).}$$

Compound interest. Here is a problem very similar to the foregoing. Suppose that a sum of money A_0 is invested at the rate of $k\%$ per year compounded continuously. To find the value A of the investment after t years, we are led to consider the differential equation

(3.5) $$\frac{dA}{dt} = \frac{k}{100} A.$$

This may best be seen as follows. If interest is earned at the rate of $k\%$ per year compounded n times a year, the amount A_n of interest plus principal at the end of t years is defined to be

$$A_n = A_0 \left(1 + \frac{k}{n} \frac{1}{100}\right)^{nt} = A_0 \left[\left(1 + \frac{k}{n} \frac{1}{100}\right)^n\right]^t.$$

To *define* the amount A at continuous compound interest, we would quite naturally use

$$A = \lim_{n \to \infty} A_n = \lim_{n \to \infty} A_0 \left[\left(1 + \frac{k}{n} \frac{1}{100}\right)^n\right]^t.$$

Since

$$\lim_{n \to \infty} \left(1 + \frac{k}{n} \frac{1}{100}\right)^n = e^{k/100},$$

we have as the amount A, which an original principal of A_0 will realize after t years, when interest is compounded continuously at the rate of $k\%$ per year,

(3.6) $A = A_0 e^{kt/100}.$

Since (3.5) and (3.6) may be regarded as equivalent statements, we see that it is appropriate to regard interest compounded continuously at the rate of $k\%$ per year as being equivalent to the idea that the rate of change of the amount is proportional to the amount present at time t, where the constant of proportionality is $\dfrac{k}{100}$.

Example. The sum of $100 is invested at the rate of 5% per year compounded continuously. When will the amount be $200?

The differential equation to be solved is

(3.5)′
$$\frac{dA}{dt} = 0.05A,$$

and its solution is

(3.6)′
$$A = 100e^{0.05t}.$$

Setting $A = 200$, we have $0.05t = \ln 2$, or $t = 13.86$ years.

Exercises

1. A tank contains 300 gal. of brine in which 20 lb. of salt is dissolved. Brine containing 0.5 lb. of salt to the gallon runs into the tank at the rate of 3 gal./min. The mixture, kept uniform by stirring, flows out at the same rate. How much salt is in solution in the tank at the end of 20 min. ?

1′. In Exercise 1, replace gallons by liters and pounds by kilograms, and solve.

2. A tank contains 50 gal. of brine in which 2 lb. of salt is dissolved. Pure salt is fed into the tank at the rate of 1 lb./min. The mixture is kept uniform and flows out of the tank at the rate of 2 gal./min. How much salt is in solution in the tank at the end of 5 min. ? What is the concentration at that time?

2′. In Exercise 2, replace gallons by liters and pounds by kilograms, and solve.

3. A tank contains 100 gal. of brine in which 5 lb. of salt is dissolved. Brine containing 0.2 lb. of salt to the gallon enters the tank at the rate of 3 gal./min. The mixture, kept uniform, flows out at the rate of 1 gal/.min. How much salt is in solution in the tank at the end of 100 min. ? What is its concentration ?

3′. In Exercise 3, replace gallons by liters and pounds by kilograms, and solve.

4.* A tank contains a l. of brine in which b kg. of salt is dissolved. Brine containing h kg. of salt to the gallon runs into the tank at the rate of g l. /min. The mixture, kept uniform by stirring, flows out of the tank at the rate of k l. /min. How much salt is in the tank at the end of t min. ? What is its concentration ? Assume $g \neq 0$, $g \neq k$.

5.   A tank contains 50 l. of a mixture containing equal parts of alcohol and water. A mixture containing 2 parts of water to 1 part of alcohol flows into the tank at the rate of 3 l./min. The fluid in the tank, kept uniform by stirring, flows out of the tank at the rate of 2 l./min. How much alcohol will there be in the solution after half an hour?

6.   A radioactive substance decomposes at a rate proportional to the amount present, and half the original quantity is left after 1,500 years. In how many years would the original amount be reduced by three-fourths? How much of the substance will be left after 2,000 years?

7.   What nominal rate of interest compounded quarterly is equivalent at the end of 1 year to 5% compounded continuously?

8.   Find the value at the end of 20 years of an investment of $100 that earns interest at the rate of 5% compounded continuously.

9.   At what nominal rate of interest per year compounded continuously must $1 be invested to double its value in 10 years?

10.   In 1626 Peter Minuit purchased Manhattan Island from the Indians for $24. If, instead, he had invested the money at 6%, compounded continuously, how much would his investment be worth 350 years later (1976)?

11. A deposits $100 at $5\frac{1}{2}$% compounded semi-annually, and B deposits $100 at 5% compounded continuously. Who is the better off, and by how much, after 5 years?

12.   A colony of bacteria increases at a rate proportional to the number present. If the number doubles in 24 hours, how many hours will be required for the bacteria to increase a hundredfold?

13.   A radioactive substance decomposes at a rate proportional to the amount present, and half the original quantity is left after 1,000 years. In how many years would the original amount be reduced by four-fifths? How much of the substance will be left after 2,000 years?

14.   In 1867 the United States purchased Alaska from Russia for $7,200,000. What would this money be worth 112 years later, if Russia had invested it at 5% interest compounded continuously?

15.   A colony of bacteria increases at a rate proportional to the number present. If the number triples in 2 hours, how long will it take for the number to be 50 times as large as it was originally?

16.   A colony of bacteria increases at a rate proportional to the number present. If the number doubles in 1 hour, how long will it take for the number to be 25 times as large as it was originally?

Answers

1.   43.6 lb.

1′.   43.6 kg.

2.   6.1 lb.; 0.30 lb./gal.

2′.   6.1 kg.; 0.30 kg./l.

3.   124.0 lb.; 0.41 lb./gal.

3′.   124.0 kg.; 0.41 kg./l.

5.   30.0 l.

6.   3,000 years; $0.4x_0$.

7.   5.03%.

8. $271.83.

9.   6.93%.

10. $31,650,000,000.

11.   *A* $131.17, *B* $128.40.

12.   160 hours.

14. $1,947,000,000.

15.   7 hours.

4   Exact differential equations of first order

A particularly important class of differential equations are the so-called *exact* differential equations. A differential equation

$$(4.1) \qquad\qquad M(x, y)\, dx + N(x, y)\, dy = 0$$

is said to be *exact* if there exists a function $g(x, y)$ such that

$$d[g(x, y)] = M(x, y)\, dx + N(x, y)\, dy;$$

that is to say, if there exists a function $g(x, y)$ such that

$$g_x(x, y) = M(x, y) \qquad \text{and} \qquad g_y(x, y) = N(x, y).$$

Thus, the equation

$$(4.2) \qquad\qquad (4x - y)\, dx + (2y - x)\, dy = 0$$

is exact, since its left-hand member is the differential of the function

$$g(x, y) = 2x^2 - xy + y^2,$$

that is,

$$d(2x^2 - xy + y^2) = (4x - y)\, dx + (2y - x)\, dy.$$

Clearly, we might equally well have chosen $g(x, y) = 2x^2 - xy + y^2 + 3$, or $g(x, y) = 2x^2 - xy + y^2 + c$, where c is any constant.

In general, of course, an expression $M(x, y)\, dx + N(x, y)\, dy$ will not be the differential of a function $g(x, y)$. For example, the expression $y\, dx - 2x\, dy$ is not exact, as we shall see later. There is no function $g(x, y)$ of which this is the differential. It is true, however, that if the functions $M(x, y)$ and $N(x, y)$

are well-behaved† in some region of the xy-plane, there always exists an *integrating factor* $I(x, y)$ such that

$$I(x, y)[M(x, y)\, dx + N(x, y)\, dy]$$

is exact. Indeed there is an infinity of integrating factors; however, except in a small number of very special cases finding one of these functions $I(x, y)$ can be very difficult. In the illustration above we may take $I(x, y) = y^{-3}$, for

$$\frac{y\, dx - 2x\, dy}{y^3} \qquad (y \neq 0)$$

is clearly the differential of x/y^2.

We come to the following definitions. When $g(x, y)$ is a differentiable function such that

$$d[g(x, y)] \equiv I(x, y)[M(x, y)\, dx + N(x, y)\, dy],$$

where $I(x, y) \neq 0$, any function $g(x, y) + c$, where c is a constant, is called an *integral* of the corresponding differential equation (4.1). Curves defined by the equations

$$g(x, y) = c \qquad (c \text{ constant})$$

are called *integral curves* of the differential equation. Accordingly, the function $2x^2 - xy + y^2$ is an integral of equation (4.2). Integral curves of equation (4.2) are given by the equation

$$(4.3) \qquad\qquad 2x^2 - xy + y^2 = c \qquad (c > 0).$$

The curves given by (4.3) are ellipses.

It is natural to inquire how we may identify those differential equations (4.1) that are exact, and how, when they are exact, corresponding integrals $g(x, y)$ may be determined. The following theorem is fundamental.

Theorem 4.1.   If the functions $M(x, y)$, $N(x, y)$ and the partial derivatives $M_y(x, y)$, $N_x(x, y)$ are continuous in a square region R, a necessary and sufficient condition that the differential equation

$$M(x, y)\, dx + N(x, y)\, dy = 0$$

be exact is that

$$(4.4) \qquad\qquad \frac{\partial M}{\partial y} \equiv \frac{\partial N}{\partial x}$$

in R.

---

† It is sufficient, for example, that $M(x, y)$, $N(x, y)$, and their partial derivatives, be continuous in a region of the xy-plane. For a proof, see the author's *An Introduction to the Theory of Ordinary Differential Equations*, Wadsworth Publishing Company (1976), p. 57.

The proof of the necessity of the condition is immediate. We suppose that the differential equation is exact; that is, there exists a function $g(x, y)$ with the property that

$$g_x(x, y) = M(x, y), \qquad g_y(x, y) = N(x, y).$$

Since $g_{xy} = g_{yx}$ it follows at once that

$$\frac{\partial M}{\partial y} = \frac{\partial N}{\partial x}.$$

The proof of the necessity is complete.

In proving the sufficiency we exhibit a function $g(x, y)$ whose partial derivatives satisfy the condition

$$g_x(x, y) = M(x, y) \qquad g_y(x, y) = N(x, y).$$

Such a function is†

$$(4.5) \qquad g(x, y) = \int_{x_0}^{x} M(x, y_0) \, dx + \int_{y_0}^{y} N(x, y) \, dy,$$

where (x_0, y_0) is a fixed point and (x, y) is an arbitrary point of the region R. For,

$$g_x = M(x, y_0) + \int_{y_0}^{y} N_x(x, y) \, dy$$

$$= M(x, y_0) + \int_{y_0}^{y} M_y(x, y) \, dy$$

$$= M(x, y_0) + [M(x, y) - M(x, y_0)]$$

$$= M(x, y),$$

while it follows at once that

$$g_y = N(x, y).$$

The proof of the theorem is complete.

Example. In the differential equation

$$(4.6) \qquad (3x^2 + y^2) \, dx + 2xy \, dy = 0$$

we see that $M(x, y) = 3x^2 + y^2$, $N(x, y) = 2xy$, and $M_y = 2y$, $N_x = 2y$.

---

† The student who is familiar with line integrals will recognize the integral in (4.5) as the line integral $\int M \, dx + N \, dy$ taken over an "elbow path" from the point (x_0, y_0) to (x, y). Condition (4.4) will then be seen to be precisely the condition that the line integral be independent of the path in R.

Some students may prefer the equivalent form

$$g(x, y) = \int_{x_0}^{x} M(t, y_0) \, dt + \int_{y_0}^{y} N(x, u) \, du.$$

Note that the variable y appears only as the upper limit of the second integral, while the variable x appears both as the upper limit of the first integral and in the integrand of the second.

Thus, the differential equation is exact. To find an integral $g(x, y)$ we choose the point (x_0, y_0) to be the origin, and we have

$$g(x, y) = \int_0^x (3x^2 + 0^2) \, dx + \int_0^y 2xy \, dy$$

$$(4.7) \qquad\qquad = x^3 + xy^2.$$

It is easily seen that the differential of (4.7) is given by the left-hand member of (4.6). Integral curves are given by the equation

$$(4.8) \qquad\qquad x^3 + xy^2 = c,$$

where c is a constant.

By finding (4.8) we have solved equation (4.6) in the sense that if we solve equation (4.8) for y and obtain a differentiable function of x, then that function is a solution of the differential equation. Specifically, we find that

$$y = \pm \sqrt{\frac{1}{x}(c - x^3)}.$$

It is easy to verify that both $\sqrt{\dfrac{1}{x}(c - x^3)}$ and $-\sqrt{\dfrac{1}{x}(c - x^3)}$ are indeed

solutions of (4.6) over a suitable interval of the x-axis.

This is the general situation, as will be seen from the following theorem.

Theorem 4.2. If $g(x, y)$ is an integral of a differential equation $M(x, y) \, dx + N(x, y) \, dy = 0$, any differentiable solution $y(x)$ of the equation $g(x, y) = c$ is a solution of the differential equation.

The proof of the theorem is omitted.

Alternate method. The integral (4.5) provides a simple and direct method of solving an exact differential equation. An alternate method, the validity of which may be established by the preceding analysis, will be illustrated by an example.

We have observed that the differential equation

$$(4.9) \qquad\qquad (3x^2 + y^2) \, dx + 2xy \, dy = 0$$

is exact. To find an integral we first integrate the term $2xy \, dy$ formally with respect to y, obtaining

$$xy^2.$$

Next, we determine a function $f(x)$, of x alone, such that

$$d[xy^2 + f(x)]$$

is given by the left-hand member of (4.9). That is, we wish to find a function $f(x)$ such that

$$2xy\, dy\, +\, y^2\, dx\, +\, f'(x)\, dx\, =\, (3x^2\, +\, y^2)\, dx\, +\, 2xy\, dy.$$

This is equivalent to the equation

$$f'(x)\, =\, 3x^2.$$

It follows that

$$f(x)\, =\, x^3,$$

and that integrals of (4.9) are given by

$$xy^2\, +\, x^3\, +\, c.$$

We might equally well have commenced by integrating formally the term $(3x^2\, +\, y^2)\, dx$, obtaining $x^3\, +\, xy^2$. We would then seek to determine a function $g(y)$, of y alone, such that $d[x^3\, +\, xy^2\, +\, g(y)]$ would be given by the left-hand member of (4.9).

An advantage of the alternate method may be observed in the first treatment of equation (4.9). Clearly, if we can determine the function $f(x)$, the equation is necessarily solved. It is desirable, however, to demonstrate that under the conditions of Theorem 4.1, such a function $f(x)$ can always be determined. This can be seen as follows. Consider the differential equation

$$M(x, y)\, dx\, +\, N(x, y)\, dy\, =\, 0,$$

and suppose that the conditions of Theorem 4.1 are satisfied. By "formal integration" of the term $N(x, y)\, dy$ is meant determining a function

$$H(x, y)\, =\, k\, +\, \int_{y_0}^{y} N(x, y)\, dy \qquad (k\ \text{constant}),$$

where the points (x, y_0) and (x, y) lie in R. We note that

$$H_y(x, y)\, =\, N(x, y),$$

(4.10) $$H_x(x, y)\, =\, \int_{y_0}^{y} N_x(x, y)\, dy\, =\, \int_{y_0}^{y} M_y(x, y)\, dy$$

$$=\, M(x, y)\Big|_{y=y_0}^{y=y}\, =\, M(x, y)\, -\, M(x, y_0).$$

To complete the demonstration we show that there exists a function $f(x)$, of x alone, such that

(4.11) $$d[H(x, y)\, +\, f(x)]\, =\, M(x, y)\, dx\, +\, N(x, y)\, dy.$$

This is equivalent to demonstrating that there exists a function $f(x)$ such that

$$H_x(x, y)\, dx + H_y(x, y)\, dy + f'(x)\, dx = M(x, y)\, dx + N(x, y)\, dy,$$

or, by (4.10), such that

$$f'(x) = M(x, y_0).$$

It is clear that $f(x)$ may be taken as

$$f(x) = \int_{x_0}^{x} M(x, y_0)\, dx,$$

if (x_0, y_0) lies in R.

Remark. It is frequently desirable in differential equation theory to note a distinction between solving a differential equation and finding a solution of a differential equation. Recall that a *solution* is always a function of x, defined on an interval, that satisfies the differential equation. On the other hand, it is customary to regard a first-order differential equation as *solved* when one can write equations of its integral curves. Theorem 4.2, of course, justifies this seeming ambiguity.

Example.   Find an integral curve of the differential equation

$$(3x^2 + y^2)\, dx + 2xy\, dy = 0$$

that passes through the point $(1, -2)$.

We saw earlier that integral curves of this differential equation are given by the equation

$$x^3 + xy^2 = c.$$

If we set $x = 1$, $y = -2$ in this equation, we find that $c = 5$; accordingly, the integral curve we seek is given by the equation

$$x^3 + xy^2 = 5.$$

Exercises

Show that the following differential equations are exact, and solve them.

1. $3x^2y\, dx + x^3\, dy = 0$. Sketch several integral curves.

2. $3(x - 1)^2\, dx - 2y\, dy = 0$.

3. $(x - y)\, dx + (2y - x)\, dy = 0$. Sketch several integral curves.

4. $(2x - y)\,dx - x\,dy = 0$. Sketch several integral curves.

5. $(x - 2y)\,dx + (4y - 2x)\,dy = 0$.

6. $\cos x \sec y\,dx + \sin x \sin y \sec^2 y\,dy = 0$.

7. $\left(x + \dfrac{y}{x^2 + y^2}\right) dx + \left(y - \dfrac{x}{x^2 + y^2}\right) dy = 0$.

8. $(y^2 + 6x^2y)\,dx + (2xy + 2x^3)\,dy = 0$.

9. $(2xy + e^y)\,dx + (x^2 + xe^y)\,dy = 0$.

10. $(2x \cos y - e^x)\,dx - x^2 \sin y\,dy = 0$.

11. $(3x^2 - 3y + e^y)\,dx + (xe^y - 3x)\,dy = 0$.

12. $(2 + 2e^y \cos 2x)\,dx + (3 + e^y \sin 2x)\,dy = 0$.

13. $(e^y \cos x - e^x \cos y)\,dx + (e^y \sin x + e^x \sin y)\,dy = 0$.

14. $\cos (x + y)(dx + dy) - \cos x \cos y\,dx + \sin x \sin y\,dy = 0$.

15. $(2xe^{2y} - y^3e^x)\,dx + (2x^2e^{2y} - 3e^xy^2)\,dy = 0$.

16. $(2xy \cos x^2 + 2 \cos y)\,dx + (\sin x^2 - 2x \sin y)\,dy = 0$.

Find the solutions $y(x)$ of the following three differential equations that satisfy the given condition.

17. $(x^2 + y^2)\,dx + 2xy\,dy = 0$, $y(1) = 1$.

18. $\dfrac{y\,dx}{x^2 + y^2} - \dfrac{x\,dy}{x^2 + y^2} = 0$, $y(2) = 2$.

19. $(x - y)\,dx + (2y - x)\,dy = 0$, $y(0) = 1$.

20.* Solve the differential equation (4.6) by evaluating the line integral $\int M\,dx + N\,dy$ along a straight line joining the points $(0, 0)$ and (x, y). (*Hint.* First, take the line integral from the origin to the point (x_1, y_1) obtaining a function $g(x_1, y_1)$. Then drop the subscripts.)

Answers

1. $x^3y = c$.

3. $x^2 - 2xy + 2y^2 = c^2$.

5. $x - 2y = c$.

9. $x^2y + xe^y = c$.

11. $x^3 - 3xy + xe^y = c$.

13. $e^y \sin x - e^x \cos y = c.$

15. $x^2 e^{2y} - y^3 e^x = c.$

17. $[(4 - x^3)/3x]^{1/2}.$

19. $\frac{1}{2}(x + \sqrt{4 - x^2}).$

5 Integrating factors

In the last section we learned how to treat an exact differential equation of the form

$$(5.1) \qquad\qquad M(x, y)\, dx + N(x, y)\, dy = 0.$$

When an equation of the form (5.1) is not exact it may be possible to find an integrating factor. Such a quantity, it will be recalled, is a function $I(x, y)$ with the property that the equation

$$(5.1)' \qquad\qquad I(x, y)[M(x, y)\, dx + N(x, y)\, dy] = 0$$

is exact. As was noted earlier, such a function, in general, exists but, except in certain special cases, it is likely to be difficult to determine.

Variables separable.   When a differential equation has the form

$$(5.2) \qquad\qquad a(x)k(y)\, dx + b(x)h(y)\, dy = 0,$$

the variables are said to be *separable*. It is easy to see that

$$\frac{1}{b(x)k(y)}$$

is an integrating factor in regions of the xy-plane in which $b(x)k(y) \neq 0$. The differential equation (5.2) can be written

$$\frac{a(x)}{b(x)}\, dx + \frac{h(y)}{k(y)}\, dy = 0,$$

which is evidently exact. Integral curves are given by

$$\int_{x_0}^{x} \frac{a(x)}{b(x)}\, dx + \int_{y_0}^{y} \frac{h(y)}{k(y)}\, dy = C.$$

These integrals will exist if, for example, the integrands are continuous on the intervals of integration.

Example.   Consider the differential equation

$$(5.3) \qquad\qquad y\, dx - x\, dy = 0.$$

It is easy to verify that any of the following functions are integrating factors:

(5.4) $$\frac{1}{xy}, \quad \frac{1}{x^2}, \quad \frac{1}{y^2}, \quad \frac{1}{x^2 + y^2},$$

except for values of x and y for which these fractions are meaningless. If we choose the first integrating factor, we have

$$\frac{dy}{y} - \frac{dx}{x} = 0,$$

and integral curves are given by

(5.5) $$\ln |y| - \ln |x| = \ln |C|,$$

or,

$$\ln \left| \frac{y}{x} \right| = \ln |C|,$$

where C is a constant. The latter is equivalent to

$$y = Cx.$$

The method we employed breaks down when $xy = 0$; that is, along the coordinate axes. A reference to equation (5.3) shows that both $x = 0$ and $y = 0$ are also integral curves. We chose the constant in (5.5) in the form $\ln |C|$ for convenience.

If we had chosen the integrating factor $\frac{1}{x^2}$, we would have been led to the equation

$$\frac{x\, dy - y\, dx}{x^2} = 0 \qquad (x \neq 0).$$

This may be rewritten as

$$d\left(\frac{y}{x}\right) = 0.$$

Integral curves are then readily seen to be given by the equation

$$\frac{y}{x} = C.$$

We must examine the curve $x = 0$ separately by reference to the original equation (5.3). Solving equation (5.3) by use of the remaining integrating factors in (5.4) is left to the reader.

We conclude this section by presenting a few situations where an integrating factor can be found by trial and error. One device which is occasionally useful may be understood from the following example.

Example.   Consider the differential equation

(5.6) $(2y^2 - 6xy)\,dx + (3xy - 4x^2)\,dy = 0.$

Here,

$$M(x, y) = 2y^2 - 6xy, \qquad N(x, y) = 3xy - 4x^2,$$

and

$$M_y = 4y - 6x, \qquad N_x = 3y - 8x.$$

Thus, equation (5.6) is not exact. We shall try to determine constants m and n so that

$$x^m y^n$$

is an integrating factor. Multiplying both members of equation (5.6) by this function we have

$$(2x^m y^{n+2} - 6x^{m+1} y^{n+1})\,dx + (3x^{m+1} y^{n+1} - 4x^{m+2} y^n)\,dy = 0.$$

This equation will be exact if (Theorem 4.1)

$$2(n + 2)x^m y^{n+1} - 6(n + 1)x^{m+1} y^n = 3(m + 1)x^m y^{n+1} - 4(m + 2)x^{m+1} y^n$$

—that is, if

$$2(n + 2)y - 6(n + 1)x = 3(m + 1)y - 4(m + 2)x,$$

identically in x and y. This will be true if

$$2(n + 2) = 3(m + 1),$$
$$6(n + 1) = 4(m + 2),$$

or if $n = 1$ and $m = 1$. Accordingly, xy is an integrating factor of (5.6).

The method used in this example can be reduced to the following test. *If there exist constants m and n such that*

$$M_y + n\,\frac{M}{y} \equiv N_x + m\,\frac{N}{x},$$

the function $x^m y^n$ is an integrating factor of (5.1).

We have seen that if

$$\frac{\partial M}{\partial y} - \frac{\partial N}{\partial x} \equiv 0,$$

the differential equation

(5.7) $M(x, y)\,dx + N(x, y)\,dy = 0$

is exact if, for example, $M(x, y)$ and $N(x, y)$ are of class C' in a rectangular (or convex) region. When

(5.8) $$\frac{1}{N}\left(\frac{\partial M}{\partial y} - \frac{\partial N}{\partial x}\right) = f(x)$$

—that is, when the left-hand member of (5.8) is a function of x alone—it can be shown that

(5.9) $$\mu(x) = e^{\int f(x)\, dx}$$

is an integrating factor for (5.7). Similarly, if

(5.10) $$\frac{1}{M}\left(\frac{\partial N}{\partial x} - \frac{\partial M}{\partial y}\right) = f(y),$$

then

(5.11) $$e^{\int f(y)\, dy}$$

is an integrating factor for (5.7).

To prove that (5.9) is an integrating factor, let us suppose that (5.7) has an integrating factor $\mu(x) \neq 0$ that is a function of x alone. Then

$$\mu M\, dx + \mu N\, dy$$

is exact, and

(5.12) $$\frac{\partial(\mu M)}{\partial y} \equiv \frac{\partial(\mu N)}{\partial x}$$

It follows that

$$\mu M_y \equiv \mu N_x + \mu' N$$

and, hence, that

(5.13) $$\frac{\mu'}{\mu} = \frac{1}{N}(M_y - N_x);$$

that is, if there is an integrating factor that is a function of x alone, the right-hand member of (5.13) must be a function $f(x)$ of x alone. We have, then, that

(5.14) $$\frac{\mu'}{\mu} = f(x),$$

and (5.9) must hold. Conversely, (5.9) leads to (5.14), and if (5.8) is assumed to hold, (5.12) follows and $\mu(x)$ is, consequently, an integrating factor for the differential equation (5.7).

The treatment of (5.10) and (5.11) is similar.

Example.   For the differential equation

(5.15) $(x^2 + y^2 + 1) \, dx - (xy + y) \, dy = 0$ $(x + 1 > 0)$,

we have

$$M_y - N_x = 3y$$

and

$$\frac{1}{N} \, (M_y - N_x) = \frac{-3}{x + 1};$$

that is,

$$f(x) = \frac{-3}{x + 1},$$

and

$$\mu(x) = e^{-\int 3 \, dx/(x + 1)} = \frac{1}{(x + 1)^3}.$$

Thus, the equation

(5.16) $$\frac{x^2 + y^2 + 1}{(x + 1)^3} \, dx - \frac{y}{(x + 1)^2} \, dy = 0$$

is exact, and integral curves of (5.15) will be found to be given by the equation

(5.17) $$\ln (x + 1) + \frac{2}{x + 1} - \frac{y^2 + 2}{2(x + 1)^2} = c,$$

where c is a constant.

Occasionally, the form of a differential equation will suggest a rearrangement that will make it exact. Consider, for example, the differential equation

(5.18) $y^2(x \, dx + y \, dy) + (x^2 + y^2)(y \, dx - x \, dy) = 0.$

We note that

$$x \, dx + y \, dy = \tfrac{1}{2} d(x^2 + y^2)$$

and that

$$d\left(\frac{x}{y}\right) = \frac{y \, dx - x \, dy}{y^2} \qquad (y \neq 0).$$

This suggests rewriting (5.18) in the form

$$\frac{x \, dx + y \, dy}{x^2 + y^2} + \frac{y \, dx - x \, dy}{y^2} = 0,$$

or

(5.19) $$\frac{d(x^2 + y^2)}{x^2 + y^2} + 2d\left(\frac{x}{y}\right) = 0.$$

Integration of (5.19) is immediate, and we have

$$\ln (x^2 + y^2) + 2\frac{x}{y} = c.$$

Problems of this sort are ordinarily in the nature of mathematical puzzles. The student will have a chance to try his hand at a few of these (in Exercises 10, 11, 13, 14, and 15 that follow).

Exercises

Separate the variables and solve the given differential equations.

1. $\sqrt{a^2 - y^2}\, dx + (b^2 + x^2)\, dy = 0.$
2. $(2x - 1)(y - 1)^2\, dx + (x - x^2)(1 + y)\, dy = 0.$
3. $x \sin y\, dx + e^{-x}\, dy = 0.$
4. $\sin y\, dx + (1 - \cos x)\, dy = 0.$
5. $y \ln x\, dx + (1 + 2y)\, dy = 0.$
6. $(2x - 1) \cos^4 y\, dx + (x^2 - 2x + 2)\, dy = 0.$
7. $y' = e^{x-y}.$
8. $e^{x+y}y' = e^{2x-y}.$
9. $e^{x^2-y^2}yy' = x + 2x^3.$

Put the following differential equations in exact form, and solve.

10. $x\, dy - y\, dx = (x^2 + y^2)\, dx.$
11. $x\, dy + y\, dx + x^4y^4(x\, dx + y\, dy) = 0.$
12. $3xy\, dx + 2x^2\, dy = 6y^3\, dx + 12xy^2\, dy.$
13. $x\, dy - y\, dx = (x^2 - y^2)\, dx.$
14. $x\, dy - y\, dx + x^2y^4(x\, dy + y\, dx) = 0.$
15. $x^2(x\, dx + y\, dy) + \sqrt{x^2 + y^2}(x\, dy - y\, dx) = 0.$

Find integrating factors for each of the following five differential equations and solve them.

16. $(x^2 + y^2)\, dx + 3xy\, dy = 0.$
17. $(x^2 + y^2)\, dx + 4xy\, dy = 0.$
18. $xy\, dx + (x^2 + 2y^2 + 2)\, dy = 0.$

19. $xy\,dx - (2x^2 + 2y^2 + 3)\,dy = 0.$

20. $3xy\,dx + (x^2 + 3y^2)\,dy = 0.$

21.   Solve the differential equation (5.16) and obtain (5.17).

Answers

1. $\dfrac{1}{b}\,\text{arc}\tan\dfrac{x}{b} + \text{arc}\sin\dfrac{y}{a} = c.$

3. $xe^x - e^x + \ln\sec y + \tan y = c.$

5. $x\ln x + \ln y + 2y = c.$

7. $e^y - e^x = c.$

9. $e^{-y^2} = (2x^2 + 3)e^{-x^2} + c.$

10. $y = x\tan(x + c).$

12. $x^3y^2 - 3x^2y^4 = c.$

14. $x^3y^4 - 3x = 3cy.$

16. $x^2(4y^2 + x^2)^3 = c.$

17. $x(x^2 + 5y^2)^2 = c.$

18. $x^2y^2 + y^4 + 2y^2 = c.$

20. $y^2(12x^2 + 9y^2)^3 = c.$

6   Homogeneous differential equations of first order

If the coefficients $M(x, y)$ and $N(x, y)$ of the differential equation

(6.1) $M(x, y)\,dx + N(x, y)\,dy = 0 \qquad [N(x, y) \not\equiv 0]$

have the property that

(6.2) $$\frac{M(ax, ay)}{N(ax, ay)} \equiv \frac{M(x, y)}{N(x, y)},$$

where a is an arbitrary number not equal to zero, the equation (6.1) is said to have *homogeneous* coefficients. When this is true, either the substitution

(6.3) $$y = vx, \qquad \frac{dy}{dx} = v + x\frac{dv}{dx},$$

or

$$(6.3)' \qquad x = wy, \qquad \frac{dx}{dy} = w + y\frac{dw}{dy}$$

will reduce (6.1) to an equation in which the variables are separable.

We shall verify this statement for the substitution (6.3). In that case (6.1) becomes

$$M(x, vx) + N(x, vx)\left(v + x\frac{dv}{dx}\right) = 0.$$

This equation may be written

$$(6.4) \qquad \frac{M(x, vx)}{N(x, vx)} + v + x\frac{dv}{dx} = 0.$$

Since equation (6.2) is an identity in the three variables a, x, and y, we may write

$$(6.2)' \qquad \frac{M(x, vx)}{N(x, vx)} \equiv \frac{M(1, v)}{N(1, v)}.$$

Accordingly, equation (6.4) may be rewritten in the form

$$[M(1, v) + vN(1, v)] + xN(1, v)\frac{dv}{dx} = 0,$$

and the variables x and v are separable.

Example. The differential equation

$$(6.1)' \qquad (x^2 + y^2)\,dx + 3xy\,dy = 0$$

has homogeneous coefficients since

$$\frac{(ax)^2 + (ay)^2}{3(ax)(ay)} \equiv \frac{x^2 + y^2}{3xy} \qquad (axy \neq 0).$$

We may employ either substitution (6.3) or (6.3)'. If the former is used, equation (6.1)' becomes

$$(x^2 + v^2x^2)\,dx + 3x(vx)(v\,dx + x\,dv) = 0,$$

or

$$\frac{dx}{x} + \frac{3v\,dv}{1 + 4v^2} = 0.$$

The formal integration is easily managed, and we have

$$\ln|x| + \tfrac{3}{8}\ln(1 + 4v^2) = \tfrac{3}{8}\ln c.$$

Setting $v = \dfrac{y}{x}$ we obtain

$$x^2(x^2 + 4y^2)^3 = c.$$

The substitution $x = wy$ would have led to a similar result.

It should be pointed out that the formal integration involved in solving such a differential equation depends on the particular substitution, (6.3) or (6.3)′, that is employed. Sometimes one leads to simpler calculations— sometimes, the other.

The student will observe that when the coefficients of a differential equation (6.1) are homogeneous, the equation can be rewritten so that its coefficients are functions of the quotient y/x (and vice versa). This is what suggests the substitution

(6.5) $\qquad \dfrac{y}{x} = v, \qquad$ or $\qquad y = xv, \qquad \dfrac{dy}{dx} = x\dfrac{dv}{dx} + v.$

Thus, equation (6.1)′ may be written as

$$\left[1 + \left(\frac{y}{x} \right)^2 \right] + 3 \left(\frac{y}{x} \right) \frac{dy}{dx} = 0.$$

The substitution (6.5) leads then to the differential equation

$$(1 + v^2) + 3v \left(x\frac{dv}{dx} + v \right) = 0,$$

and solving this equation proceeds as above.

In general, we might then rewrite the differential equation (6.1) in the form

$$f\left(\frac{y}{x} \right) + g\left(\frac{y}{x} \right) \frac{dy}{dx} = 0.$$

The substitution (6.5) leads to the equation

$$f(v) + g(v)\left(x\frac{dv}{dx} + v \right) = 0,$$

or

$$[f(v) + vg(v)]\, dx + xg(v)\, dv = 0.$$

The variables are separable, for the last equation may be rewritten as

$$\frac{dx}{x} + \frac{g(v)\, dv}{f(v) + vg(v)} = 0.$$

An integrating factor. We note, in passing, that when the equation

$$M(x, y)\, dx + N(x, y)\, dy = 0$$

has homogeneous coefficients, an integrating factor may be shown to be

$$\frac{1}{xM + yN}.$$

The substitutions (6.3) and (6.3)$'$ ordinarily lead to simpler computations than does the use of this integrating factor.

Occasionally a differential equation with coefficients that are almost homogeneous can be rewritten in homogeneous form by use of simple transformations. For example, consider the differential equation

(6.6) $$(x - y - 1)\, dx + (x + 4y - 6)\, dy = 0.$$

The coefficients in this differential equation would be homogeneous if the constant terms -1 and -6 were not present. This situation can be remedied by setting

$$x = u + h, \qquad y = v + k,$$

where h and k are constants to be determined. Noting that $dx = du$ and $dy = dv$, we have

(6.7) $$[u - v + (h - k - 1)]\, du + [u + 4v + (h + 4k - 6)]\, dv = 0.$$

We attempt to choose h and k so that

$$h - k - 1 = 0,$$

$$h + 4k - 6 = 0.$$

When these simultaneous equations are solved, we find that $h = 2$, $k = 1$, and equation (6.7) becomes

(6.8) $$(u - v)\, du + (u + 4v)\, dv = 0,$$

which has homogeneous coefficients. This equation may then be solved by means of the substitution

$$u = wv, \qquad du = w\, dv + v\, dw.$$

After the usual computation, integral curves of equation (6.8) are found to be

$$\ln (u^2 + 4v^2) = \arctan\left(\frac{u}{2v}\right) + C.$$

We now set $u = x - 2$, $v = y - 1$, and obtain

$$\ln\left[(x - 2)^2 + 4(y - 1)^2\right] = \arctan\left[\frac{x - 2}{2(y - 1)}\right] + C \qquad (y \neq 1)$$

as an equation of integral curves of the given differential equation (5.9).

Exercises

Show that the first ten differential equations have homogeneous co-efficients and solve.

1. $(x - y)\,dx + x\,dy = 0$. (Use both (6.3) and (6.3)′; also solve as a linear differential equation.)

2. $(x - 2y)\,dx + x\,dy = 0$. (Use both (6.3) and (6.3)′.)

3. $(x^2 - y^2)\,dx + 2xy\,dy = 0$.

4. $\sqrt{x^2 + y^2}\,dx = x\,dy - y\,dx$.

5. $(x^2y + 2xy^2 - y^3)\,dx - (2y^3 - xy^2 + x^3)\,dy = 0$.

6. $\left(x \sin\dfrac{y}{x} - y \cos\dfrac{y}{x}\right)dx + x \cos\dfrac{y}{x}\,dy = 0$.

7. $(x^3 + 2xy^2)\,dx + (y^3 + 2x^2y)\,dy = 0$. Check your answer by solving as an exact differential equation.

8. $[x \sin(e^{y/x}) - ye^{y/x} \cos(e^{y/x})]\,dx + xe^{y/x} \cos(e^{y/x})\,dy = 0$.

9. $\left(x^2 \sin\dfrac{y^2}{x^2} - 2y^2 \cos\dfrac{y^2}{x^2}\right)dx + 2xy \cos\dfrac{y^2}{x^2}\,dy = 0$.

10. $(x^2 e^{-y^2/x^2} - y^2)\,dx + xy\,dy = 0$.

Solve the following differential equations.

11. $(2x + y - 2)\,dx + (2y - x + 1)\,dy = 0$.

12. $(x + y - 10)\,dx + (x - y - 2)\,dy = 0$. Check your answer by solving as an exact differential equation.

13. $(x - 3y)\,dx + (x + y + 4)\,dy = 0$.

14. $(x + 3y)\,dx + (3x - 2y + 5)\,dy = 0$.

15. $(x - y)\,dx + (x - y + 2)\,dy = 0$. (*Hint.* Set $u = x - y$).

16. $(x + 2y + 1)\,dx + (3x + 6y + 2)\,dy = 0$.

17. $(x^2 + y^2)\,dx + kxy\,dy = 0$ (k constant).

18. $(x^2 + 4xy + y^2)\,dx + (y^2 + 2xy + 2x^2)\,dy = 0$. Also, solve as an exact differential equation.

19. $(x^2 - y^2)\,dx + (4y^2 - 2xy)\,dy = 0$. Also, solve as an exact differential equation.

20. $(x^2 + 6xy - 12y^2)\,dx + (3x^2 - 24xy)\,dy = 0$. Also, solve as an exact differential equation.

21. $(x^2 + 2xy - 6y^2)\,dx + (x^2 - 12xy)\,dy = 0$. Also, solve as an exact differential equation.

22. $\left(xy\cos\dfrac{x}{y} + 2y^2\sin\dfrac{x}{y}\right)dx - \left[(x^2 + y^2)\cos\dfrac{x}{y} + xy\sin\dfrac{x}{y}\right]dy = 0.$

23. $\left(x - y\ln\dfrac{y}{x}\right)dx + x\ln\dfrac{y}{x}\,dy = 0.$

24. $[x^2(1 + e^{y^2/x^2}) - 2y^2 e^{y^2/x^2}]\,dx + 2xye^{y^2/x^2}\,dy = 0.$

25.*   Solve the differential equation
$$(x^2 + 4xy^2 - y^4)\,dx + (4y^5 + 4xy^3 - 4x^2 y)\,dy = 0.$$
(*Hint.* Determine constants m and n so that the substitution $x = u^m$, $y = v^n$ reduces the equation to homogeneous form.)

Answers

1. $xe^{y/x} = c.$

3. $x^2 + y^2 = 2ax.$

5. $(x^2 - y^2)e^{x/y} = c.$

9. $x\sin(y^2/x^2) = c.$

11. $\ln[(x - 1)^2 + y^2] - \arctan[y/(x - 1)] = c.$

13. $\ln|x - y + 2| + \dfrac{2y + 2}{x - y + 2} = c.$

15. $x + y - \ln|x - y + 1| = c.$

17. $x[x^2 + (k + 1)y^2]^{k/2} = c$ $(k \neq -1)$; $y^2 = x^2\ln c^2 x^2$ $(k = -1).$

19. $4y^3 - 3xy^2 + x^3 = c.$

21. $x^3 + 3x^2 y - 18xy^2 = c.$

23. $\dfrac{y}{x}\left(\ln\left|\dfrac{y}{x}\right| - 1\right) + \ln|x| = c.$

25.   Any pair of numbers that satisfies the equation $m = 2n$, except $(0, 0)$, will do; $x^2 + y^4 = c(x + 2y^2).$

7   Orthogonal trajectories

The equation

(7.1) $$x^2 + y^2 = c \qquad (c > 0 \text{ constant})$$

represents a circle with center at the origin and radius $\sqrt{c}$. If c is regarded as a parameter, (7.1) may be interpreted as an equation of a family of circles with centers at the origin and of variable radius $\sqrt{c}$. If both members of (7.1) are differentiated with respect to x, the constant c is eliminated, and we have the differential equation

(7.2) $$x + yy' = 0,$$

or

(7.2)' $$x\,dx + y\,dy = 0.$$

It is readily seen, conversely, that equation (7.1) represents integral curves of this differential equation. The family (7.1) is called a *one-parameter* family since it depends on a single parameter c.

Note that there is one and only one member of the family (7.1) which passes through each point of the plane except the origin (see Fig. 1.1). For

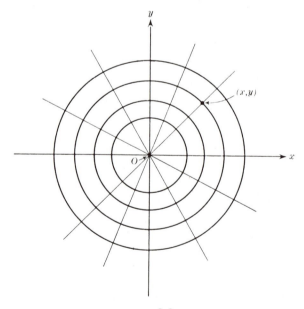

FIG. 1.1

example, the circle through the point $(-3, 4)$ has the equation

$$x^2 + y^2 = 25;$$

that is, $c = 25$. In general, if the circle passes through the point (x_0, y_0), then $c = x_0^2 + y_0^2$, and the circle of the family through this point has the equation

$$x^2 + y^2 = c \qquad (c = x_0^2 + y_0^2).$$

At the point (x, y) the slope of the circle passing through this point is given by equation (7.2):

$$(7.7) \qquad\qquad y' = -\frac{x}{y} \qquad (y \neq 0).$$

Suppose we wish to find a second family of curves with the property that each member of the second family intersects each member of the first family at right angles, or *orthogonally*. When this occurs each family of curves is called *orthogonal trajectories* of the other. Because the slope of each circle of the given family at the point (x, y) is given by $-\frac{x}{y}$, the slope of an orthogonal trajectory to the circle at this point must be the negative reciprocal of this number, or $\frac{y}{x}$. That is, in the family of curves which cut the circles orthogonally we must have

$$y' = \frac{y}{x} \qquad (x \neq 0),$$

or

$$(7.4) \qquad\qquad x\,dy - y\,dx = 0.$$

The variables are separable in equation (7.4). We see at once then that among the integral curves associated with (7.4) are the straight lines

$$(7.5) \qquad\qquad y = cx \qquad (c \neq 0).$$

This is a family of straight lines passing through the origin. To avoid dividing by zero we required both $x \neq 0$ and $y \neq 0$ in the preceding analysis. It is clear from the geometry (see Fig. 1.1), however, that the orthogonal trajectories of the family of circles (7.1) are the family (7.5) plus the lines $x = 0$ and $y = 0$.

This suggests a general method. Suppose the equation

$$(7.6) \qquad\qquad f(x, y) = c$$

represents a one parameter family F_1 of curves. If $f(x, y)$ is differentiable, the curves (7.6) are integral curves of the differential equation

$$f_x(x, y)\,dx + f_y(x, y)\,dy = 0.$$

The orthogonal trajectories of the family (7.6) are then integral curves F_2 of the differential equation (see Fig. 1.2)

$$f_y(x, y)\,dx - f_x(x, y)\,dy = 0.$$

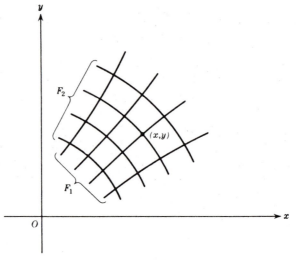

FIG. 1.2

Exercises

1. Find a first-order differential equation possessing the given family of integral curves:

    (a) $y = cx$;

    (b) $y^2 = 2ax$;

    (c) $x^2 + y^2 - 2ax = 0$;

    (d) $xy = c$.

2. Find the orthogonal trajectories of the one-parameter family of curves given below. Sketch a few members of each family.

    (a) $y = cx$;

    (b) $y^2 = 2ax$;

    (c) $x^2 + y^2 - 2ax = 0$;

    (d) $x^2 + 4y^2 = 4a^2$;

    (e) $xy = c$;

    (f) $y = ce^x$;

    (g) $x^2 - 2y^2 = c$;

    (h) $a^2x^2 + y^2 = a^2$.

Answers

1. (a) $x\,dy - y\,dx = 0$; (c) $(x^2 - y^2)\,dx + 2xy\,dy = 0$.

2. (a) $x^2 + y^2 = a^2$; (c) $x^2 + y^2 - 2cy = 0$; (f) $y^2 + 2x = c$;
    (h) $y^2 = \ln x^2 - x^2 + c$.

8 Equations reducible to differential equations of first order

Consider the second-order differential equation

$$(8.1) \qquad y'' - y' = 0$$

This equation is of first order in the variable y'. This observation may be exploited by the use of the substitution

$$(8.2) \qquad p = y'.$$

Equation (8.1) then becomes the first-order differential equation

$$(8.1)' \qquad p' - p = 0$$

the solution of which is

$$p = c_1 e^x \qquad (c_1 \text{ constant}).$$

Thus, we set

$$y' = c_1 e^x,$$

and an integration yields

$$y = c_1 e^x + c_2 \qquad (c_2 \text{ constant}).$$

The substitution (8.2) will ordinarily be effective when the letter y is not present in the given differential equation. When the letter x is not present, a helpful substitution frequently is

$$y' = p,$$

$$(8.3)$$

$$y'' = \frac{dp}{dx} = \frac{dp}{dy}\frac{dy}{dx} = p\frac{dp}{dy}.$$

Example.   Consider the differential equation

$$(8.4) \qquad yy'' = 1 + y'^2.$$

We note that the variable x does not appear formally, and we employ the substitution (8.3), obtaining

$$\frac{2p\,dp}{1 + p^2} = 2\frac{dy}{y}.$$

Integrating, we have

$$\ln(1 + p^2) = \ln y^2 + \ln c^2,$$

or†

$$1 + p^2 = c^2 y^2.$$

---

† Note that the constant must be positive.

Thus,

$$p = y' = \pm \sqrt{c^2 y^2 - 1},$$

and

(8.5)
$$\frac{dy}{\sqrt{c^2 y^2 - 1}} = \pm dx.$$

Integration of (8.5) may be accomplished by the substitution

$$cy = \sec \theta, \qquad c\,dy = \sec \theta \tan \theta \, d\theta,$$

and we have

$$\frac{1}{c} \ln |cy + \sqrt{c^2 y^2 - 1}| = \pm (x - c) + c_1.$$

Exercises

Solve the following differential equations.

1. $y'' - 2y' = 0$ (two methods).

2. $y'' + y' = 0$ (two methods).

3. $xy'' - y' = 0$.

4. $xy'' + 3y' = 0$.

5. $y'' + y'^2 + 1 = 0$.

6. $yy'' + y'^2 + 1 = 0$.

7. $y'' = 2yy'$.

8. $y'' = (1 + y'^2)^{3/2}$. [*Note.* This equation may also be solved by inspection.]

9. $2yy'' = 1 + y'^2$.

10. $y'' + 4y = 0$.

11. $y'' + 9y = 0$.

12. $y'' - 4y = 0$.

13. $y'' - 9y = 0$.

14. Solve the differential equation

$$y'' + a^2 y = 0$$

by multiplying through by $2y'$ and integrating.

15.* Note Exercise 14 and show that a solution $y(x)$ of the differential equation

$$y'' + p(x)y = 0$$

satisfies the identity

$$y'^2(x) + p(x)y^2(x) \equiv c + \int_a^x y^2 p'(x)\, dx,$$

if $p'(x)$ is continuous on an interval $a \le x \le b$.

16.* Show that integrals of the differential equation

$$(1 - x^2 - y^2)y'' + 2(1 + y'^2)(xy' - y) = 0$$

are provided by the family of circles

$$(x - a)^2 + (y - b)^2 = a^2 + b^2 - 1.$$

The center of each circle is the variable point (a, b), and its radius is $\sqrt{a^2 + b^2 - 1}$. It is an interesting problem to characterize this family geometrically. Can you do it?

Answers

1. $y = c_1 + c_2 e^{2x}$.

3. $y = c_1 + c_2 x^2$.

5. $y = \ln \sin (x + c_1) + c_2$.

7. $y = c \tan (cx + k)$.

9. $4k(y - k) = (x - h)^2$.

11. $a \sin 3x + b \cos 3x$.

13. $ae^{3x} + be^{-3x}$.

Supplementary Exercises

Solve the given differential equations.

1. $(x - y)\, dx + (x + y)\, dy = 0$.

2. $xy' + 2y = x^2 - 2x + 3$.

3. $(x \cot y - 2 \cos y)\, dy + dx = 0$.

4. $(\sin y + y \sec^2 x)\, dx + (\tan x + x \cos y)\, dy = 0$.

5. $(3y - x)\, dx + (3x + 2y)\, dy = 0$.

6. $(1 - x^2 + y^2 - x^2 y^2)\, dx + (x \arctan y)\, dy = 0$.

7. $(x^2 + y^2)\, dx - 2xy\, dy = 0.$

8. $\dfrac{dy}{dx} = \dfrac{6x - y - 3}{3x + 2y + 1}.$

9. $(x^2 - 3y^2 + 2)\, dx + (2xy + 4x^4)\, dy = 0.$

10. $y(e^{\frac{x}{y}} + 1)\, dx + [y(e^{\frac{x}{y}} - 1) - xe^{\frac{x}{y}}]\, dy = 0.$

11. $(x - 2y)\, dx + (6y - 3x + 1)\, dy = 0.$

12. $xy' + y = -x(1 + x^2)y^2.$

13. $xy' + y = (3 - 2x)y^{-2}.$

   Solve the following differential equations by substituting $y = vx^n$, where n is a suitably chosen constant.

14. $(x^4 + y^2)\, dx - xy\, dy = 0.$

15.* $(2x^3 + 4xy^3)\, dx + (6x^2y^2 - 3y^5)\, dy = 0.$

16.* $(2x^5y + 4x^3y^2 - 2xy^3)\, dx + (x^2y^2 - 2y^3 - x^6)\, dy = 0.$

Answers

1. $\ln (x^2 + y^2) + 2 \arctan \dfrac{y}{x} = c.$

3. $2x \sin y + \cos 2y = c.$

5. $2y^2 + 6xy - x^2 = c.$

7. $y^2 = x^2 - cx$

9. $y^2 - x^2 + \frac{2}{3} = x^3(c - 4y).$

11. $3y - x - \ln |1 - x + 2y| = c.$

13. $[3(1 - \frac{1}{2}x + cx^{-3})]^{\frac{1}{3}}.$

15. $n = \frac{2}{3};\ x^4 + 4x^2y^3 - y^6 = c.$

2

The existence and nature
of solutions

1 Preliminary remarks

In the preceding chapter we were able to find solutions of certain special differential equations. The methods employed were tailored to take advantage of the peculiarities of form exhibited by the differential equation being studied. We cannot go very far, nor can we acquire more than superficial understanding, simply by treating such special cases. For example, the differential equation

$$(2x + y^2)\, dx + 2xy\, dy = 0$$

is readily seen to be exact, and it may be solved by methods familiar to us. A slight change in the coefficients yields the equation

$$(1.1) \qquad (2x + y^2)\, dx + 3xy\, dy = 0,$$

which is no longer exact, and a different method must be sought.

We may ask, indeed, if there exists a solution of (1.1), and, if so, under what conditions. Are there general methods for finding it which do not involve special devices geared to the special peculiarities of form the equation possesses? Can we tell something about the behavior of a solution without solving the differential equation? In this chapter we begin our study of such questions.

Recall that there are algebraic equations which do not possess solutions. For example, the equation

$$x^2 + 1 = 0$$

has no *real* solution x. Similarly, it is easy to construct examples of differential equations that have no solution. There is no *real* function that satisfies the differential equation

$$y'^2 + y^2 + 1 = 0.$$

And other less obvious examples can be constructed.

Before we can discuss the existence of solutions of differential equations we shall need to define a solution with some care.

Definition. A solution of the nth-order differential equation

(1.2) $$y^{(n)} = f[x, y, y', \ldots, y^{(n-1)}]$$

is a function $y(x)$ defined on an interval of the x-axis such that

$$y^{(n)}(x) \equiv f[x, y(x), y'(x), \ldots, y^{(n-1)}(x)]$$

on that interval.

This definition implies that a solution is of at least class C^{n-1} and that it possesses an nth derivative on this interval.† Note that a solution is a function of x defined on an interval. When f is a continuous function of its $n + 1$ variables, a solution is, of course, of class C^n on its interval; that is, it possesses a continuous nth derivative on the interval.

It follows from the definition that a solution of a first-order differential equation

$$y' = f(x, y)$$

must be continuous on its interval I of definition and possess a derivative at each point of I. Similarly, a solution of a second-order differential equation

$$y'' = f(x, y, y')$$

---

† A function $f(x)$ which, together with its first k derivatives, is continuous on an interval is said to be of class C^k $(k = 0, 1, 2, \ldots)$ on the interval. A function of class C^0 is, then, simply a continuous function.

must be continuous on its interval I of definition. Further, it must possess a continuous first derivative on I and it must also have a second derivative at each point of the interval.

To illustrate the definition consider the first-order differential equation

$$y' = 2x + 1.$$

The function $x^2 + x$ is clearly a solution of this equation over every interval on the x-axis. So is the function $x^2 + x + c$, where c is any constant.

Again, the second-order differential equation

(1.3) $$y'' + y = 0$$

has the solution $\sin x$ on every interval of the x-axis. Another solution of this differential equation is $\cos x$. Indeed, every function $c_1 \sin x + c_2 \cos x$, where c_1 and c_2 are constants, is a solution of this differential equation on every interval.

Finally, observe that the differential equation

$$y' = 2x + 1 \qquad (0 \le x < 1)$$

has the solution $x^2 + x + c$ on the interval $0 \le x < 1$. Note that the differential equation has been assigned no meaning outside this interval.

Close scrutiny of the following examples is recommended. To understand what is *not* a solution is helpful in comprehending what a solution *is*.

Example. The differential equation

(1.4) $$y' = 1$$

has the solution $y = x - 1$ over every interval. It also has the solution $y = x + 1$ over every interval. The function $y(x)$ defined on the interval $-1 \le x \le 1$ by the equations

$$y(x) = x - 1 \qquad (-1 \le x \le 0),$$
$$= x + 1 \qquad (0 < x \le 1),$$

however, is *not* a solution of (1.4) on the interval $[-1, 1]$, for it is not continuous at $x = 0$.

Example. Consider the differential equation

(1.5) $$y'' = 0.$$

A solution on every interval is clearly the function $-x$. So, also, is $+x$. The function $y(x)$ defined on the interval $-1 \le x \le 1$ by the equations

(1.6)
$$y(x) = -x \qquad (-1 \le x < 0),$$
$$= x \qquad (0 \le x \le 1),$$

although continuous on this interval, is *not* a solution on this interval. Since equation (1.5) is of second order, a solution on an interval must, by definition, be a function with a continuous derivative on that interval that possesses a second derivative at each point of the interval. The curve $y = y(x)$ defined by equations (1.6) has a corner at $x = 0$ and thus $y(x)$ does not have a first derivative at that point.

Next, let us consider the following problem. Find a solution $y(x)$ of the differential equation

$$(1.7) \qquad\qquad y' = e^x$$

such that $y(0) = 0$. A correct answer is

$$(1.8) \qquad y(x) = \int_0^x e^x \, dx \qquad (-\infty < x < \infty),$$

or, since it does not matter what letter is used for the variable of integration,

$$(1.9) \qquad y(x) = \int_0^x e^t \, dt \qquad (-\infty < x < \infty).$$

Our problem may be regarded as solved since it is clear from either (1.8) or (1.9) that $y'(x) = e^x$ and that $y(0) = 0$. Note that if a solution $y(x)$ of (1.7) were required having the property that $y = 1$ when $x = 0$ [that is, $y(0) = 1$], a correct answer would be

$$(1.10) \qquad\qquad y(x) = 1 + \int_0^x e^t \, dt.$$

It is, in a sense, accidental and rather fortunate that we may, if we choose, rewrite the solution (1.8) and (1.9) as

$$(1.11) \qquad\qquad y(x) = e^x - 1.$$

A solution of the differential equation

$$y' = e^{x^2},$$

which has the property that $y = 0$ when $x = 0$, is, similarly,

$$(1.12) \qquad y(x) = \int_0^x e^{t^2} \, dt \qquad (-\infty < x < \infty).$$

Equation (1.12) constitutes a complete and valid answer to our problem, even though we cannot carry out the indicated formal integration in the ordinary elementary sense. For further information concerning the nature of the function defined by equation (1.12), we might resort to one of various approximate methods—perhaps Simpson's rule, or power series expansions.

It should be emphasized, however, that any difficulties we may encounter in approximating the integral in question affect not at all the validity of the answer (1.12). Indeed, even to evaluate the function $e^x - 1$ in (1.11) when $x = 2$, for example, leads us just as surely to approximate methods. Fortunately, the approximate values of e^x have been tabulated in a number of places, and so we feel a kind of relative familiarity with this function.

Exercises

1. Find the solution of the differential equation $y' = 1 + y^2$ that satisfies the condition $y(0) = 0$. What is the largest interval of validity for your solution?

2. Do the same as in Exercise 1, when the differential equation is $y' + y = x$, and $y(0) = 1$.

3. Do the same as in Exercise 1, when the differential equation is

$$xy' + y = \cos x \ (x > 0), \text{ and } y(\pi) = 0.$$

4. Do the same as in Exercise 1, when the differential equation is

$$y' = -(1 + y^2), \text{ and } y(\pi/2) = 0.$$

In the following exercises you are given a differential equation and its general solution involving constants c_1 and c_2. Find the particular solution that satisfies the given conditions.

5. $y'' + y = 0$, $c_1 \sin x + c_2 \cos x$, $y(0) = 1$, $y'(0) = 0$.

6. $y'' - y = 0$, $c_1 e^x + c_2 e^{-x}$, $y(0) = 1$, $y'(0) = 1$.

7. $x^2 y'' - 2xy' + 2y = 0$, $c_1 x + c_2 x^2$, $y(1) = 0$, $y'(1) = 1$.

8. $y'' + a^2 y = 0$, $c_1 \sin ax + c_2 \cos ax$, $y(0) = \alpha$, $y'(0) = \beta$.

9. $x^2 y'' - xy' + 5y = 0$, $c_1 x \sin (2 \ln x) + c_2 x \cos (2 \ln x)$, $y(1) = y'(1) = 1$.

10. $y'' + y = 0$, $c_1 \sin x + c_2 \cos x$, $y(0) = 2$, $y'(\pi) = 1$.

11. $y'' - y = 0$, $c_1 \sinh x + c_2 \cosh x$, $y(0) = 1$, $y(1) = 0$.

12. $y'' - 4y = 0$, $c_1 \sinh 2x + c_2 \cosh 2x$, $y(0) = 0$, $y(2) = 0$.

Answers

1. $\tan x \left(-\dfrac{\pi}{2} < x < \dfrac{\pi}{2} \right)$.

3. $\dfrac{\sin x}{x} \ (x > 0)$.

5. $\cos x$.

7. $x^2 - x$.

9. $x \cos (2 \ln x)$.

11. $\dfrac{1 + e^2}{1 - e^2} \sinh x + \cosh x$.

2   Direction fields

Consider the first-order differential equation

(2.1)
$$\frac{dy}{dx} = -\frac{1}{4}\frac{x}{y}.$$

Viewing this differential equation geometrically we observe that it provides the slope of the integral curve through the point (x, y) at that point. That is, (2.1) provides what is called a *direction field* for the xy-plane. The curves cross the y-axis ($x = 0$) with zero slope and intersect the x-axis ($y = 0$) vertically. It is also often helpful to note the curves that the integral curves intersect with constant slope k—the so-called *isoclines* of the differential equation. In (2.1), the isoclines are the straight lines

$$y = -\frac{1}{4k} x.$$

We observe, for example, that the integral curves have slope 1 where they intersect the line $y = -x/4$ (see Fig. 2.1).

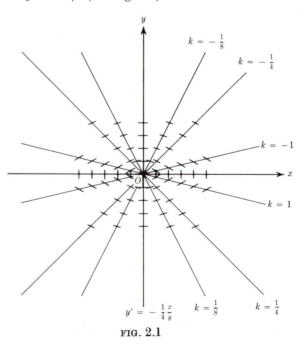

FIG. 2.1

Such observations frequently enable us to obtain useful rough pictures of the family of integral curves associated with a differential equation of the form

(2.2) $$\frac{dy}{dx} = f(x, y).$$

Equation (2.1) has, of course, relatively simple integral curves. They are the family of ellipses (Figure 2.2)

$$x^2 + 4y^2 = c.$$

The student should compare Figs. 2.1 and 2.2.

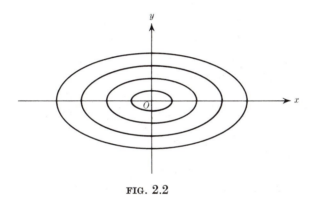

FIG. 2.2

A slightly more complicated example is provided by the differential equation

(2.3) $$\frac{dy}{dx} = x + y.$$

In this case the isoclines are the straight lines

(2.4) $$x + y = k.$$

This is the family of parallel lines of slope -1. The integral curves of (2.3) meet a line (2.4) with slope k. Figure 2.3 displays the isoclines for (2.3), elements of its direction field, and several integral curves. Solving the linear differential equation (2.3) yields the fact that the integral curves have the equation $y = -x - 1 + ce^x$.

As we shall see in the next section, there is precisely one integral curve through each point of the xy-plane.

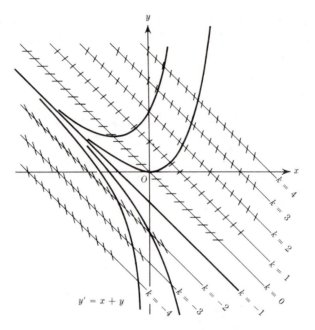

FIG. 2.3

Exercises

In the following exercises sketch the isoclines, elements of the direction field, and four or five integral curves of the given differential equation.

1. $y' = x/y.$

2. $y' = x - y.$

3. $y' = \frac{1}{2}y - x.$

4. $y' = -y/x.$

5. $y' = x + 2y.$

6. $y' = xy.$

Answers

1. Isoclines: $y = \frac{1}{k}x$; integral curves: $y^2 - x^2 = c.$

3. Isoclines: $y - 2x = k$; integral curves: $y = 2x + 4 + ce^{x/2}.$

5. Isoclines: $x + 2y = k$; integral curves: $y = -\frac{1}{2}x - \frac{1}{4} + ce^{2x}.$

3　The fundamental existence theorem

In this section we give statements† of the fundamental existence theorem that may be readily applied. We shall first state a form of the theorem as it concerns a differential equation of first order.

Theorem 3.1.　Let $f(x, y)$ and $f_y(x, y)$ be continuous functions of the variables x and y in a region‡ R of the xy-plane containing the point (x_0, y_0). If a constant h is taken positive and sufficiently small, there exists one and only one solution $y(x)$ of the differential equation

$$(3.1) \qquad\qquad y' = f(x, y)$$

on the interval $x_0 \le x \le x_0 + h$ with the property that $y(x_0) = y_0$.

The following examples will clarify the use of the theorem.

Example.　Consider the *differential system*

$$(3.2) \qquad\qquad y' = y$$

$$(3.3) \qquad\qquad y(0) = 1.$$

Here, we seek a solution $y(x)$ of equation (3.2) that satisfies condition (3.3). In this example (x_0, y_0) is the point $(0, 1)$, and

$$f(x, y) = y,$$
$$f_y(x, y) = 1.$$

This function $f(x, y)$ clearly satisfies the hypotheses of the theorem throughout the xy-plane. In particular, we may apply the theorem in any region containing the point $(0, 1)$. According to Theorem 3.1, there exists one and only one solution $y(x)$ of (3.2) satisfying (3.3). We note by inspection that e^x is a solution of (3.2). Further, it satisfies (3.3). Since there is only one such solution, according to Theorem 3.1, e^x is the solution we seek.

If the point had been $(0, 0)$, the unique solution through this point would have been the *null* solution $y(x) = 0$.

The following examples illustrate what may happen when the hypotheses of the theorem do not hold.

Examples.　Consider the differential equation

$$(3.4) \qquad\qquad y' = -\frac{1}{x} y.$$

---

† For a proof of the fundamental existence theorem see the author's *An Introduction to the Theory of Ordinary Differential Equations*, Wadsworth Publishing Company, Belmont, California (1976).

‡ By a *region* is meant an open, connected set.

Here the hypotheses of the theorem fail to hold for the right-hand member of (3.4) along the line $x = 0$. Every solution of (3.4) has the form $y = c/x$, where c is a constant (see Fig. 2.4). Thus, all solutions except the null solution become infinite as x tends to zero. Note also that there is no solution through any point $(0, a)$ $(a \neq 0)$. If $x_0 \neq 0$, the unique solution of (3.4) through the point (x_0, y_0)† is seen to be

$$y(x) = x_0 y_0 x^{-1}$$

on any interval not containing the origin.

A similar equation is the differential equation

(3.5) $$xy' = y.$$

All solutions of this equation are of the form $y = cx$, where c is a constant. It will be seen that again no solution goes through the point $(0, a)(a \neq 0)$, whereas all solutions pass through the origin. Through every point (x_0, y_0) $(x_0 \neq 0)$, however, there passes exactly one solution $y(x)$ of (3.5). This solution is readily calculated to be $y(x) = x_0^{-1} y_0 x$.

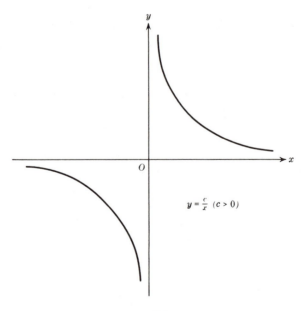

FIG. 2.4

---

† By "solution through a point (x_0, y_0)" is meant a solution $y(x)$ with the property that the curve $y = y(x)$ passes through the point (x_0, y_0).

Next, we state a form of the existence theorem as it applies to a differential equation of second order. The alterations necessary for the extension to differential equations of higher order will be clear.

Theorem 3.2. Let $f(x, y, z), f_y(x, y, z)$, and $f_z(x, y, z)$ be continuous functions of the variables x, y, z in some three-dimensional region R containing the point (x_0, y_0, z_0). If h is taken positive and sufficiently small, there exists one and only one solution $y(x)$ of the differential equation

$$y'' = f(x, y, y')$$

on the interval $x_0 \leq x \leq x_0 + h$ such that $y(x_0) = y_0$ and $y'(x_0) = z_0$.

Theorems 3.1 and 3.2 remain valid if in their statements the positive constant h is replaced by a positive constant h' and the interval is taken as $x_0 - h' \leq x \leq x_0$. We have then the following extension of Theorem 3.2 (an analogous extension of Theorem 3.1 is readily stated).

Theorem 3.3. Let $f(x, y, z)$, $f_y(x, y, z)$, $f_z(x, y, z)$ be continuous functions of the variables x, y, z in some three-dimensional region R containing the point (x_0, y_0, z_0). If h' and h are taken positive and sufficiently small, there exists one and only one solution $y(x)$ of the differential equation

$$y'' = f(x, y, y')$$

on the interval $x_0 - h' \leq x \leq x_0 + h$ such that $y(x_0) = y_0$ and $y'(x_0) = z_0$.

The following examples will illustrate Theorem 3.3.

Example. Consider the differential system

$$y'' + y = 0,$$

$$y(0) = 0,$$

$$y'(0) = 2.$$

In this case, $f(x, y, z) = -y$, and the point (x_0, y_0, z_0) is the point $(0, 0, 2)$. Since $f_y(x, y, z) = -1$, and $f_z(x, y, z) = 0$, the conditions of Theorem 3.3 are satisfied for all points (x, y, z). Note that $2 \sin x$ is a solution of the differential system. According to the fundamental existence theorem (Theorem 3.3), there is only one solution having this property; therefore, the solution we seek is $2 \sin x$.

Example.   Consider the differential system

$$y'' + 2xy' + x^{-2}y = 0,$$

(3.6) $$y(1) = 0,$$

$$y'(1) = 0.$$

In this case $f(x, y, z)$ is the function $-2xz - x^{-2}y$, for $f(x, y, y')$ then becomes $-2xy' - x^{-2}y$. Here,

$$f_y(x, y, z) = -x^{-2}, \qquad f_z = -2x,$$

and so the conditions of Theorem 3.3 are met in any region in which $x \neq 0$. From (3.6) we observe that the point (x_0, y_0, z_0) of the theorem is, in this instance, $(1, 0, 0)$, and Theorem 3.3 guarantees a unique solution of the system (3.6). Note that $y(x) \equiv 0$ is, first, a solution of the differential equation (obvious by inspection) and, second, a solution such that $y(1) = 0$ and $y'(1) = 0$. According to Theorem 3.3 there is only one solution having this property. Hence, the solution we seek is the function 0.

The student will observe that, roughly speaking, the existence theorem states that for a differential equation

$$y^{(n)} = f(x, y, y', \ldots, y^{(n-1)})$$

one may prescribe values for $y, y', \ldots, y^{(n-1)}$ at a point $x = x_0$ and be assured of the existence of a unique solution of the differential equation satisfying these conditions.

We state next the fundamental existence theorem as it applies to the *general nth-order linear differential equation*

(3.7) $$a_0(x)y^{(n)} + a_1(x)y^{(n-1)} + \cdots + a_{n-1}(x)y' + a_n(x)y = f(x),$$

where $a_0(x) \neq 0$, and $a_0(x), a_1(x), \ldots, a_n(x), f(x)$ are continuous on an interval I.

Theorem 3.4. Let $x = x_0$ be a point of the interval I and let $c_0, c_1, c_2, \ldots, c_{n-1}$ be an arbitrary set of n real numbers. There then exists one and only one solution $y(x)$ of (3.7) with the property that

$$y(x_0) = c_0, \qquad y'(x_0) = c_1, \qquad \ldots, \qquad y^{(n-1)}(x_0) = c_{n-1}.$$

Further, this solution is defined over the entire interval I.

Thus, at any point $x = x_0$ of the interval we may prescribe the values $y(x_0), y'(x_0), \ldots, y^{(n-1)}(x_0)$ quite arbitrarily, and there will then exist precisely one solution over the whole interval that, together with its first $n - 1$ derivatives, assumes the prescribed values at the point $x = x_0$.

For example, suppose the interval I is the interval $0 \leq x \leq 1$ and that $n = 3$. Then at the point $x = \frac{1}{3}$ we may prescribe that $y = \sqrt{2}$, $y' = \pi^2$, $y'' = 10^{10}$, and the theorem asserts the existence of one and exactly one solution $y(x)$ taking on these values at $x = \frac{1}{3}$. This solution will, further, be defined at every point of the interval $0 \leq x \leq 1$, and $y(x)$, $y'(x)$, $y''(x)$, and $y'''(x)$ will be continuous functions at each point of this interval.

Recall that a function $f(x)$ that, together with its first n derivatives, is continuous on an interval is said to be of class C^n on this interval. Thus, a solution of (3.7) is by definition of class C^{n-1}, and it follows from (3.7) that it is also of class C^n on I.

It is important to note that if we discover (no matter how) a solution $y(x)$ on I that satisfies the conditions

$$y(x_0) = c_0, \qquad y'(x_0) = c_1, \qquad \ldots, \qquad y^{(n-1)}(x_0) = c_{n-1},$$

we have found *the* solution we seek. For $y(x)$ will obviously be *a* solution of our problem, but by Theorem 3.4 there is exactly one such solution; hence, $y(x)$ is *the* solution.

It should be noted that, unlike the general situation, the solution of a *linear* differential equation guaranteed by the fundamental existence theorem is valid on the entire interval I. The differential equation

$$y' = 1 + y^2$$

is *not* linear and has the unique solution $y(x) = \tan x$ that satisfies the condition $y(0) = 0$. Observe that the maximum interval of existence of this solution is $-\pi/2 < x < \pi/2$. On the other hand, the linear differential equation

$$y' = 1 + y$$

has the unique solution $y(x) = e^x - 1$ that satisfies the condition $y(0) = 0$. This solution exists on the interval $-\infty < x < \infty$.

If $f(x) \equiv 0$ on I, equation (3.7) becomes

(3.8) $\qquad a_0(x)y^{(n)} + a_1(x)y^{(n-1)} + \cdots + a_{n-1}(x)y' + a_n(x)y = 0.$

We say then that (3.8) is *homogeneous*.

Corollary 1. *A solution $y(x)$ of (3.8) with the property that at a point $x = x_0$ of I*

(3.9) $\qquad y(x_0) = 0, \qquad y'(x_0) = 0, \qquad \ldots, \qquad y^{(n-1)}(x_0) = 0$

is identically zero on I.

The proof of the corollary is extremely simple. We note that $y(x) = 0$ is first of all *a* solution of (3.8) that satisfies (3.9). By the fundamental existence theorem it is the *only* solution with this property. The proof is complete.

The student should particularly note the following special case of Corollary 1.

Corollary 2. If a solution of a first-order linear homogeneous differential equation

$$a_0(x)y' + a_1(x)y = 0$$

vanishes at a single point of I, the solution is identically zero on I.

We are assuming, as usual, that the functions $a_0(x)$ and $a_1(x)$ are continuous and $a_0(x) \neq 0$ on I.

The student will readily verify that if $y_1(x)$ and $y_2(x)$ are any two solutions of (3.8), the *linear combination* $c_1 y_1(x) + c_2 y_2(x)$, where c_1 and c_2 are constants, is also a solution of (3.8). In the same fashion it can readily be seen that a linear combination of any number of solutions of (3.8) is a solution of (3.8).

Exercises

1. Given the differential equation

$$y'' + a(x)y' + b(x)y = 0,$$

where $a(x)$ and $b(x)$ are continuous on the interval $a \leq x \leq b$, what can be said of a solution $y(x)$ whose graph $y = y(x)$ is tangent to the x-axis at a point of the interval?

2. Find the solution of the differential system

$$y' = 1 - y^2,$$
$$y(0) = 0,$$

on the interval $-\infty < x < \infty$.

3. Find the unique solution $y(x)$ of the differential system

$$(2x - y)\, dx + (2y - x)\, dy = 0,$$
$$y(1) = 1.$$

Over what interval of the x-axis is this function a solution?

4. Find the unique solution $y(x)$ of the differential system

$$x\, dy - y\, dx - (x^2 + y^2)\, dx = 0,$$
$$y\left(\frac{\pi}{2}\right) = \frac{\pi}{2}.$$

5. Find the unique solution $y(x)$ of the differential system

$$xy' + y = 2x,$$
$$y(1) = 2.$$

Over what interval of the x-axis is this function a solution?

6. Find the unique solution $y(x)$ of the differential system

$$(x^2 - y^2)\, dx + 2xy\, dy = 0,$$
$$y(1) = 1.$$

Over what interval of the x-axis is this function a solution?

7. Find the unique solution $y(x)$ of the differential system

$$xy'' + 2y' = 0,$$
$$y(-1) = 0,$$
$$y'(-1) = 2.$$

Over what interval of the x-axis is this function a solution?

8. Find the unique solution $y(x)$ of the differential system (see Exercise 8, Section 1)

$$y'' + a^2y = 0 \qquad (a > 0),$$
$$y(0) = 0,$$
$$y'(0) = a.$$

9. Find the unique solution $y(x)$ of the differential system (see Exercise 7, Section 1)

$$x^2y'' - 2xy' + 2y = 0,$$
$$y(1) = 1,$$
$$y'(1) = 4.$$

10. Find the unique solution $y(x)$ of the differential system (see Exercise 9, Section 1).

$$x^2y'' - xy' + 5y = 0,$$
$$y(1) = 1,$$
$$y'(1) = 2.$$

11. Given that every solution of the differential equation $y'' + y = 0$ can be written in the form $c_1 \sin x + c_2 \cos x$, where c_1 and c_2 are constants, decide whether or not there exist solutions of the following differential systems. When a solution exists, is it unique?

(a) $y'' + y = 0$,
    $y(0) = 1$,
    $y'(0) = 0$.

(d) $y'' + y = 0$,
    $y(-\pi) = 0$,
    $y(0) = 0$.

(b) $y'' + y = 0$,
    $y(0) = 0$,
    $y(\pi) = 1$.

(e) $y'' + y = 0$,
    $y(0) = 2$,
    $y''(0) = -2$.

(c) $y'' + y = 0$,
    $y(0) = 1$,
    $y(\pi) = -1$.

12. Note Exercise 11 and show that every solution $y(x)$ of the differential system

$$y'' + y = 0,$$
$$y(0) = 0,$$

has the property that $y(\pi) = 0$.

13. Given the differential equation of Exercise 1, what can be said of two solutions which are tangent to each other at a point (x_0, y_0)?

14. By analogy, state an existence theorem for a differential equation of the form

$$y''' = f(x, y, y', y'').$$

Answers

1. $y(x) \equiv 0$.

2. $\tanh x$.

3. $\frac{1}{2}(x + \sqrt{4 - 3x^2})$; $\frac{-2}{\sqrt{3}} < x < \frac{2}{\sqrt{3}}$.

5. $x + \frac{1}{x}$; $0 < x < \infty$.

7. $-2 - \frac{2}{x}$; $-\infty < x < 0$.

9. $3x^2 - 2x$.

11. (a) $\cos x$ (uniquely).
    (b) No solution.
    (e) $2 \cos x + c_1 \sin x$.

3

Linear differential equations
with constant coefficients

1 Second-order equations

The differential equation

(1.1) $a(x)y'' + b(x)y' + c(x)y = f(x)$ $[a(x) \neq 0]$,

where $a(x)$, $b(x)$, $c(x)$, and $f(x)$ are continuous on an interval I [recall (2.3) of Chapter 1, Section 2], is the general linear differential equation of second order. According to the existence theorem [Chapter 2, Section 3], if $x = x_0$ is any point of I and c_0 and c_1 are any two numbers, there exists one and only one solution $y(x)$ of (1.1) such that

$$y(x_0) = c_0 \quad \text{and} \quad y'(x_0) = c_1.$$

Further, this solution is defined on the entire interval I.

The differential equation

$$(1.2) \qquad a(x)y'' + b(x)y' + c(x)y = 0$$

is called the *associated, or corresponding, homogeneous* equation of (1.1). Because $a(x) \neq 0$, one may divide through both equations (1.1) and (1.2) by $a(x)$ obtaining equations of the form

$$(1.1)' \qquad y'' + b(x)y' + c(x)y = f(x),$$

$$(1.2)' \qquad y'' + b(x)y' + c(x)y = 0,$$

respectively. The functions $b(x), c(x)$, and $f(x)$ in (1.1)$'$ and (1.2)$'$ are, of course, not the functions $b(x)$, $c(x)$, and $f(x)$ of (1.1) and (1.2), unless $a(x)$ is the function 1.

We shall deal first with the *homogeneous* differential equations

$$(1.3) \qquad y'' + b(x)y' + c(x)y = 0,$$

where $b(x)$ and $c(x)$ are continuous on an interval I.

An example of (1.3) is

$$(1.4) \qquad y'' + y = 0.$$

Here, I may be taken to be the interval $-\infty < x < \infty$. Note that $\sin x$ is a solution of this equation. So is $\cos x$; indeed, so is the function

$$(1.5) \qquad c_1 \sin x + c_2 \cos x,$$

where c_1 and c_2 are any constants.

Another example is

$$(1.6) \qquad y'' - \frac{2}{x}y' + \frac{2}{x^2}y = 0.$$

Here, I may be taken to be the interval $0 < x < \infty$, or, in fact, any interval of the x-axis that does not contain the origin. A solution on any such interval is the function x; so is x^2, and so is the function

$$(1.7) \qquad c_1 x + c_2 x^2,$$

where c_1 and c_2 are any constants.

We shall see shortly that (1.5) and (1.7) are *general solutions* of their respective differential equations; that is, every such function is a solution, and every solution can be written in that form.

The notions of *linear dependence* and *linear independence* of solutions are fundamental in the study of linear differential equations. Two functions $u(x)$ and $v(x)$ are said to be *linearly dependent*, if there exist constants c_1 and c_2, not both zero, such that

$$(1.8) \qquad c_1 u(x) + c_2 v(x) \equiv 0 \qquad (\text{on } I).$$

Otherwise—that is, if the identity (1.8) implies that $c_1 = c_2 = 0$, the solutions are said to be *linearly independent*.

To illustrate the idea of linear independence, let us first show that the solutions x and x^2 of equation (1.6) are linearly independent on $I: 0 < x < \infty$. Suppose there are constants c_1 and c_2 such that

$$(1.9) \qquad c_1 x + c_2 x^2 \equiv 0 \qquad (\text{on } I).$$

Then the derivative of the left-hand member of this identity must also be zero on I; that is,

$$(1.10) \qquad c_1 + 2c_2 x \equiv 0 \qquad (\text{on } I).$$

In particular, (1.9) and (1.10) must then hold when $x = 1$; thus,

$$c_1 + c_2 = 0,$$
$$c_1 + 2c_2 = 0.$$

It follows at once that $c_1 = c_2 = 0$ and that the functions x and x^2 are linearly independent on I. The point $x = 1$ of I was chosen for convenience. Any other point of I would have done as well.

Let us apply the definition to the solutions $\sin x$ and $\cos x$ of equation (1.4). Suppose there are constants c_1 and c_2 such that

$$(1.11) \qquad c_1 \sin x + c_2 \cos x \equiv 0 \qquad (I: -\infty < x < \infty).$$

Then, as above, the derivative of the left-hand member of this identity must also be zero, and we have

$$(1.12) \qquad c_1 \cos x - c_2 \sin x \equiv 0.$$

Again, (1.11) and (1.12) must hold when $x = 0$; that is,

$$c_2 = 0,$$
$$c_1 = 0,$$

and we see that $\sin x$ and $\cos x$ are linearly independent functions on I. The point $x = 0$ of I was chosen for convenience. Any other point of I would lead to the same conclusion.

Note, however, that $2 \sin x$ and $3 \sin x$ are linearly dependent solutions, for

$$3(2 \sin x) + (-2)(3 \sin x) \equiv 0.$$

Here, $c_1 = 3$, $c_2 = -2$.

We now turn our attention to some general properties of solutions of equation (1.3). First, we note the following.

1. *If $u(x)$ is a solution of (1.3), so is $ku(x)$, where k is any constant.*
2. *If $u(x)$ and $v(x)$ are any two solutions of (1.3), so is $u(x) + v(x)$.*

To prove the first statement, we have given that

(1.13) $$u''(x) + b(x)u'(x) + c(x)u(x) \equiv 0 \qquad \text{(on } I\text{)}.$$

Substituting $ku(x)$ in the left-hand side of (1.3) we have

$$[ku(x)]'' + b(x)[ku(x)]' + c(x)[ku(x)],$$

which we wish to show is identically zero. But this can be written as

$$k[u''(x) + b(x)u'(x) + c(x)u(x)],$$

which by (1.13) is identically zero, and the proof is complete.

The proof of the second statement is just as easy and is left to the student.

Note that if $u(x)$ and $v(x)$ are solutions, so is $c_1 u(x) + c_2 v(x)$, where c_1 and c_2 are any constants. For, by the first statement, $c_1 u(x)$ and $c_2 v(x)$ are both solutions, and, by the second, so is their sum.

A basic theorem that we shall employ is the following.

Theorem 1.1. If the coefficients $b(x)$ and $c(x)$ of the differential equation

$$y'' + b(x)y' + c(x)y = 0$$

are continuous on an interval I, there exist infinitely many pairs of solutions linearly independent on I. If $u(x)$ and $v(x)$ are any such pair, the general solution of the differential equation can be written in the form

(1.14) $$c_1 u(x) + c_2 v(x),$$

where c_1 and c_2 are constants.

Further, all solutions are valid on the entire interval I.

We have seen that $\sin x$ and $\cos x$ are linearly independent solutions of the differential equation

$$y'' + y = 0.$$

It follows from the theorem that the general solution of this equation is

(1.15) $c_1 \sin x + c_2 \cos x,$

where c_1 and c_2 are constants—that is, every solution of this differential equation can be put in this form and every such function (1.15) is a solution on I.

We have already observed that if $u(x)$ and $v(x)$ are any two solutions of an equation (1.3), so is $c_1 u(x) + c_2 v(x)$, where c_1 and c_2 are constants, and that all solutions of a linear differential equation are always valid on all of I. To prove the rest of the theorem, let g, h, k, and m be any four constants such that the determinant

(1.16) $\begin{vmatrix} g & h \\ k & m \end{vmatrix} \neq 0.$

Pick any point $x = x_0$ of I. By the fundamental existence theorem, there exist two solutions $u(x)$ and $v(x)$ of (1.3) such that

$$u(x_0) = g, \qquad v(x_0) = h,$$
$$u'(x_0) = k, \qquad v'(x_0) = m.$$

We shall first prove that $u(x)$ and $v(x)$ are linearly independent. To that end, suppose that there exist constants c_1 and c_2 such that

(1.17) $c_1 u(x) + c_2 v(x) \equiv 0 \qquad$ (on I).

Then, also,

(1.18) $c_1 u'(x) + c_2 v'(x) \equiv 0 \qquad$ (on I).

The identities (1.17) and (1.18) must hold, in particular, when $x = x_0$; that is,

$$c_1 u(x_0) + c_2 v(x_0) = 0,$$
$$c_1 u'(x_0) + c_2 v'(x_0) = 0,$$

or

$$c_1 g + c_2 h = 0,$$
$$c_1 k + c_2 m = 0.$$

But these are simultaneous equations in c_1 and c_2 with a nonzero determinant. Therefore, $c_1 = c_2 = 0$, and the solutions $u(x)$ and $v(x)$ are linearly independent.

It remains to be shown that every solution of (1.3) can be written as a linear combination of $u(x)$ and $v(x)$. That is, if $y(x)$ is an arbitrary solution of (1.3), there exist constants c_1 and c_2 such that

$$y(x) \equiv c_1 u(x) + c_2 v(x).$$

Let c_1 and c_2 be the constants determined by the equations

(1.19)
$$c_1 u(x_0) + c_2 v(x_0) = y(x_0),$$
$$c_1 u'(x_0) + c_2 v'(x_0) = y'(x_0),$$

or, what is the same thing,

$$c_1 g + c_2 h = y(x_0),$$
$$c_1 k + c_2 m = y'(x_0).$$

These constants exist and are uniquely determined because of condition (1.16). For this choice of constants consider the solution

$$w(x) = y(x) - c_1 u(x) - c_2 v(x)$$

of (1.3). We see that $w(x_0) = 0$ and $w'(x_0) = 0$, by (1.19). It follows that $w(x) \equiv 0$, and the theorem is proved.

Exercises

1. Use the argument of this section to show that the following pairs of functions are linearly independent on the given interval.

    (a) e^x, e^{-x} $(-\infty < x < \infty)$; (b) x^2, x^3 $(0 < x < \infty)$;
    (c) $\cosh x, \sinh x$ $(-\infty < x < \infty)$; (d) $x \sin x, x \cos x$ $(0 < x < \infty)$;
    (e) $x \sin x^2, x \cos x^2$ $(0 < x < \infty)$; (f) $\sin 2x, \cos 2x$ $(-\infty < x < \infty)$;
    (g) e^{2x}, e^{-3x} $(-\infty < x < \infty)$; (h) xe^x, e^x $(-\infty < x < \infty)$;
    (i) $e^x \sin x, e^x \cos x$ $(-\infty < x < \infty)$; (j) $(1 + x^2) \cosh x, (1 + x^2) \sinh x$
    $\qquad\qquad\qquad\qquad\qquad\qquad\qquad (-\infty < x < \infty)$.

    [*Note.* In this exercise we are not concerned with the fact that the above pairs are solutions of a differential equation of the form (1.3).]

2. Prove that if $u(x)$ and $v(x)$ are solutions of (1.3) so is $u(x) + v(x)$.

3. Show that if $u(x)$ and $v(x)$ are solutions of (1.3) such that, for any point $x = x_0$ of I,

    $$\begin{vmatrix} u(x_0) & v(x_0) \\ u'(x_0) & v'(x_0) \end{vmatrix} = 0,$$

    the solutions are linearly dependent.

2 Constant coefficients

The *wronskian* of two solutions $u(x)$ and $v(x)$ of the differential equation

$$y'' + b(x)y' + c(x)y = 0,$$

is the determinant

$$\begin{vmatrix} u(x) & v(x) \\ u'(x) & v'(x) \end{vmatrix} = u(x)v'(x) - u'(x)v(x).$$

Later we shall show that a necessary and sufficient condition that solutions $u(x)$ and $v(x)$ be linearly dependent is that their wronskian vanish identically on I. Meanwhile, we shall use this result.

To illustrate, consider the differential equation

$$(2.1) \qquad\qquad y'' - y = 0 \qquad (-\infty < x < \infty).$$

It is easy to verify that e^{-x} and e^x are solutions of this differential equation. Their wronskian is the determinant

$$\begin{vmatrix} e^{-x} & e^x \\ -e^{-x} & e^x \end{vmatrix} = 2;$$

consequently, these solutions are linearly independent, and the general solution of (2.1) is then

$$c_1 e^{-x} + c_2 e^x,$$

where c_1 and c_2 are constants.

The functions $\cosh x$ and $\sinh x$ are also solutions of (2.1), as is readily verified. They are linearly independent inasmuch as their wronskian

$$\begin{vmatrix} \cosh x & \sinh x \\ \sinh x & \cosh x \end{vmatrix} = \cosh^2 x - \sinh^2 x \equiv 1.$$

Accordingly, the general solution of (2.1) may also be written in the form

$$c_1 \cosh x + c_2 \sinh x.$$

Consider now the differential equation

$$(2.2) \qquad\qquad ay'' + by' + cy = 0 \qquad (a \neq 0),$$

where a, b, and c are (real) constants. To solve such an equation it is natural to recall that the general solution of the first-order linear differential equation

$$y' + ay = 0$$

is given by

$$ce^{-ax} \qquad (c \text{ constant}).$$

Perhaps, then, there is a solution of (2.2) of the form

$$e^{mx} \qquad (m \text{ constant}).$$

Substituting e^{mx} for y in (2.2) we have

$$e^{mx}(am^2 + bm + c) = 0.$$

That is, e^{mx} will be a solution of (2.2) provided m is a root of the *characteristic* (or *indicial*) equation

(2.3) $$am^2 + bm + c = 0.$$

Equation (2.3) is a quadratic equation, and three cases arise:

 1. The roots m_1 and m_2 are real and unequal.
 2. The roots m_1 and m_2 are real and equal.
 3. The roots m_1 and m_2 are conjugate complex numbers.

The roots of equation (2.3) are called *characteristic* (or *indicial*) *roots*.

Case 1: roots real and unequal.   Consider the differential equation

(2.2)' $$y'' + y' - 2y = 0.$$

The characteristic equation for this differential equation is

(2.3)' $$m^2 + m - 2 = 0,$$

the roots of which are $m_1 = 1$, $m_2 = -2$. Thus, solutions of (2.2)' are

$$e^x \qquad \text{and} \qquad e^{-2x}.$$

Their wronskian is

$$\begin{vmatrix} e^{-2x} & e^x \\ -2e^{-2x} & e^x \end{vmatrix} = 3e^{-x} \neq 0 \qquad (-\infty < x < \infty).$$

It follows that the general solution of (2.2)' is

$$c_1 e^x + c_2 e^{-2x}.$$

In general, when the roots m_1 and m_2 of (2.3) are real and unequal, solutions of (2.2) are

$$e^{m_1 x} \qquad \text{and} \qquad e^{m_2 x}.$$

Their wronskian is

$$\begin{vmatrix} e^{m_1 x} & e^{m_2 x} \\ m_1 e^{m_1 x} & m_2 e^{m_2 x} \end{vmatrix} = (m_2 - m_1)e^{(m_1 + m_2)x} \neq 0,$$

and the general solution of (2.2) is

$$c_1 e^{m_1 x} + c_2 e^{m_2 x}.$$

Case 2: roots real and equal.   Consider the differential equation

(2.2)'' $$y'' - 2y' + y = 0 \qquad (-\infty < x < \infty).$$

Its characteristic equation is

$$(2.3)'' \qquad\qquad m^2 - 2m + 1 = 0$$

the roots of which are $m_1 = m_2 = 1$. Thus,

$$e^x$$

is a solution of $(2.2)''$. In order to write down the general solution of this differential equation, we need to find a second linearly independent solution. The characteristic roots have provided only one solution.

A second solution is easily verified [by substitution in $(2.2)''$] to be the function†

$$xe^x.$$

The wronskian of these two solutions is

$$\begin{vmatrix} e^x & xe^x \\ e^x & xe^x + e^x \end{vmatrix} = e^{2x} \neq 0,$$

and the general solution of $(2.2)''$ is then

$$c_1 e^x + c_2 xe^x.$$

We shall see that the example above is typical for when the roots of the quadratic equation (2.3) are real and equal; that is, when

$$b^2 - 4ac = 0,$$

and the repeated root is

$$(2.4) \qquad\qquad m = -\frac{b}{2a}.$$

Accordingly, one solution of (2.2) is then

$$e^{-(b/2a)x}.$$

To show that xe^{mx} is also a solution set

$$y = xe^{mx}, \qquad y' = e^{mx}(mx + 1), \qquad y'' = e^{mx}(m^2 x + 2m)$$

in (2.2), and we have

$$e^{mx}[a(m^2 x + 2m) + b(mx + 1) + cx] = e^{mx}[x(am^2 + bm + c) + (2am + b)].$$

---

† This solution can be derived by setting $y = ve^x$ in (2.1) and determining a function $v(x)$ so that $v(x)e^x$ is a solution.

But the coefficient of x inside the brackets is zero, because m is a characteristic root. The remaining quantity vanishes because of (2.4). Thus, xe^{mx} is indeed a solution, and the wronskian of these two solutions is

$$\begin{vmatrix} e^{mx} & xe^{mx} \\ me^{mx} & (mx + 1)e^{mx} \end{vmatrix} = e^{2mx} \neq 0.$$

The solutions are then linearly independent, and the general solution is, accordingly,

$$c_1 e^{mx} + c_2 x e^{mx}.$$

Case 3: roots conjugate complex numbers. Consider the differential equation

(2.5) $$y'' - 4y' + 13y = 0.$$

Its associated indicial equation is

$$m^2 - 4m + 13 = 0,$$

the roots of which are $m_1 = 2 + 3i$, $m_2 = 2 - 3i$. We shall see that linearly independent solutions of (2.4) are

$$e^{2x} \cos 3x \qquad \text{and} \qquad e^{2x} \sin 3x.$$

If we set

$$y = e^{2x} \cos 3x, \qquad y' = e^{2x}(2 \cos 3x - 3 \sin 3x),$$
$$y'' = e^{2x}(-5 \cos 3x - 12 \sin 3x),$$

in (2.5), we see that the differential equation is satisfied. In the same way, it may be seen that $e^{2x} \sin 3x$ is also a solution. The wronskian of these two solutions is

$$\begin{vmatrix} e^{2x} \cos 3x & e^{2x} \sin 3x \\ e^{2x}(2 \cos 3x - 3 \sin 3x) & e^{2x}(2 \sin 3x + 3 \cos 3x) \end{vmatrix} = 3e^{4x} \neq 0;$$

consequently, the solutions are linearly independent, and the general solution of (2.5) is

$$c_1 e^{2x} \cos 3x + c_2 e^{2x} \sin 3x.$$

These solutions arise from some elementary manipulations of complex numbers. The indicial roots $2 + 3i$, $2 - 3i$ lead formally to the solutions

(2.6) $$e^{(2 + 3i)x} \qquad \text{and} \qquad e^{(2 - 3i)x}.$$

It is proved in the theory of functions of a complex variable that

$$\frac{d}{dx}\, e^{(\alpha + i\beta)x} = (\alpha + i\beta)e^{(\alpha + i\beta)x},$$

$$\frac{d^2}{dx^2}\, e^{(\alpha + i\beta)x} = (\alpha + i\beta)^2 e^{(\alpha + i\beta)x}.$$

Accordingly, the functions (2.6) may be readily seen to be formal (complex) solutions of (2.5).

Next, a famous formula, due to Euler, is this:

$$(2.7) \qquad\qquad\qquad e^{u + iv} = e^u(\cos v + i \sin v).$$

The solutions (2.6) may then be written, respectively, as

$$e^{2x}(\cos 3x + i \sin 3x),$$

$$(2.8)$$

$$e^{2x}(\cos 3x - i \sin 3x).$$

Again, even when the solutions of such an equation are complex, the sum and difference of two solutions are solutions, as is a constant times a solution. If we add the solutions (2.8) and subtract them, we obtain, respectively, the solutions

$$(2.9) \qquad\qquad 2e^{2x} \cos 3x \qquad \text{and} \qquad 2ie^{2x} \sin 3x.$$

Finally, we multiply these solutions by the constants $1/2$ and $1/2i$, respectively, to obtain the solutions we seek.

The above solution of (2.5) represents the general situation. Suppose the roots of the quadratic equation

$$am^2 + bm + c = 0 \qquad (a > 0)$$

are $m_1 = \alpha + i\beta$, $m_2 = \alpha - i\beta$, where α and β are real, and $\beta > 0$. Then,

$$\alpha = -\frac{b}{2a}, \qquad \beta = \frac{\sqrt{4ac - b^2}}{2a} \qquad (4ac - b^2 > 0).$$

The functions (2.8) are replaced by

$$e^{\alpha x}(\cos \beta x + i \sin \beta x),$$

$$(2.8)'$$

$$e^{\alpha x}(\cos \beta x - i \sin \beta x),$$

while (2.9) becomes

$$(2.9)' \qquad\qquad 2e^{\alpha x} \cos \beta x \qquad \text{and} \qquad 2ie^{\alpha x} \sin \beta x.$$

Multiplying these solutions by the appropriate constants, we obtain the solutions

$$e^{\alpha x} \cos \beta x \qquad \text{and} \qquad e^{\alpha x} \sin \beta x$$

of the differential equation.

It remains to show that these solutions are always linearly independent. To that end, compute their wronskian

$$\begin{vmatrix} e^{\alpha x} \cos \beta x & e^{\alpha x} \sin \beta x \\ e^{\alpha x}(\alpha \cos \beta x - \beta \sin \beta x) & e^{\alpha x}(\beta \cos \beta x + \alpha \sin \beta x) \end{vmatrix} = 2\beta e^{2\alpha x} \neq 0.$$

The general solution is then

$$c_1 e^{\alpha x} \cos \beta x + c_2 e^{\alpha x} \sin \beta x.$$

Example.   Find the solution $y(x)$ of the differential equation

$$y'' - 4y' + 13y = 0$$

that satisfies the conditions

$$y(0) = 0, \qquad y'(0) = 1.$$

This is the differential equation (2.5), and its general solution has been found to be

$$y(x) = c_1 e^{2x} \cos 3x + c_2 e^{2x} \sin 3x.$$

We seek to determine constants c_1 and c_2 such that $y(0) = 0$, $y'(0) = 1$. Differentiating, we have

$$y'(x) = c_1 e^{2x}(2 \cos 3x - 3 \sin 3x) + c_2 e^{2x}(3 \cos 3x + 2 \sin 3x).$$

The given conditions yield

$$0 = y(0) = c_1,$$
$$1 = y'(0) = 2c_1 + 3c_2.$$

It follows that $c_1 = 0$, $c_2 = 1/3$, and the solution we seek is

$$\tfrac{1}{3} e^{2x} \sin 3x.$$

Exercises

Show that the given functions are solutions of the given differential equation. By computing their wronskian, prove that they are linearly independent solutions.

1. $y'' + y = 0$, $\sin x, \cos x$ $(-\infty < x < \infty)$.

2. $y'' + y = 0$, $\sin x, \sin x - \cos x$ $(-\infty < x < \infty)$.

3. $y'' - y = 0,$ e^x, e^{-x} $(-\infty < x < \infty).$

4. $y'' - 4y = 0,$ $\cosh 2x, \sinh 2x$ $(-\infty < x < \infty).$

5. $y'' - 2y' + 2y = 0,$ $e^x \cos x, e^x \sin x$ $(-\infty < x < \infty).$

6. $y'' - 8y' + 16y = 0,$ e^{4x}, xe^{4x} $(-\infty < x < \infty).$

Find two linearly independent solutions of the following differential equations and write down their general solutions.

7. $y'' - 3y' + 2y = 0.$

8. $y'' + 2y' + 2y = 0.$

9. $y'' + 4y' + 4y = 0.$

10. $y'' + 4y' + 8y = 0.$

11. $y'' + 6y' + 9y = 0.$

12. $y'' - 4y' + 3y = 0.$

13. $8y'' + 4y' + y = 0.$

14. $4y'' + 4y' + y = 0.$

15. $y'' - 6y' + 13y = 0.$

16. $y'' - 2y' + 5y = 0.$

In Exercises 17–24, find the general solution of the differential equation. Then obtain the solution determined by the given conditions.

17. $y'' - 2y' - 3y = 0, y(0) = 0, y'(0) = 1.$

18. $y'' - y' = 0, y(0) = 2, y'(0) = 1.$

19. $y'' + 6y' + 13y = 0, y(0) = 1, y'(0) = -1.$

20. $y'' + 6y' - 12y = 0, y(0) = 0, y'(0) = 0.$

21. $y'' - 2y' - y = 0, y(0) = 0, y'(0) = \sqrt{2}.$

22. $y'' + 2y' = 0, y(0) = -3, y'(0) = 10.$

23. $y'' - 2y' + 3y = 0, y(0) = 1, y'(0) = 0.$

24. $y'' + 4y' + 13y = 0, y(0) = 0, y'(0) = -1.$

Find the general solutions of the following differential equations.

25. $y'' + a^2 y = 0$ (a constant, $\neq 0$).

26. $y'' = 0.$

27. $y'' - 2\alpha y' + (\alpha^2 + \beta^2)y = 0$ (α, β constants, $\beta \neq 0$).

28. $y'' - 2ay' + (a^2 - c^2)y = 0$ (a, c constants, $c > 0$).

Answers

7. $c_1 e^x + c_2 e^{2x}$.

8. $e^{-x}(c_1 \cos x + c_2 \sin x)$.

9. $c_1 e^{-2x} + c_2 x e^{-2x}$.

11. $c_1 e^{-3x} + c_2 x e^{-3x}$.

13. $e^{-x/4}(c_1 \cos \frac{1}{4}x + c_2 \sin \frac{1}{4}x)$.

15. $c_1 e^{3x} \cos 2x + c_2 e^{3x} \sin 2x$.

17. $c_1 e^{-x} + c_2 e^{3x}; \frac{1}{4}(e^{3x} - e^{-x})$.

19. $e^{-3x}(c_1 \cos 2x + c_2 \sin 2x); e^{-3x}(\cos 2x + \sin 2x)$.

21. $c_1 e^{(1+\sqrt{2})x} + c_2 e^{(1-\sqrt{2})x}; \frac{1}{2}(e^{(1+\sqrt{2})x} - e^{(1-\sqrt{2})x})$.

23. $e^x[c_1 \cos (\sqrt{2}x) + c_2 \sin (\sqrt{2}x)]; e^x[\cos (\sqrt{2}x) - \dfrac{\sqrt{2}}{2} \sin (\sqrt{2}x)]$.

25. $c_1 \cos ax + c_2 \sin ax$.

27. $c_1 e^{\alpha x} \cos \beta x + c_1 e^{\alpha x} \sin \beta x$.

28. $c_1 e^{(a+c)x} + c_2 e^{(a-c)x}$.

3   The inhomogeneous second-order linear differential equation

Consider the differential equation

$$(3.1) \qquad a(x)y'' + b(x)y' + c(x)y = f(x) \qquad [a(x) \neq 0],$$

where $a(x)$, $b(x)$, $c(x)$, and $f(x)$ are continuous, with $a(x) \neq 0$, on some interval I. When $f(x) \not\equiv 0$, equation (3.1) is called *inhomogeneous* (or *nonhomogeneous*). The basic theorem about such differential equations is the following.

Theorem 3.1.   If $y_0(x)$ is any particular solution of the differential equation

$$(3.2) \qquad a(x)y'' + b(x)y' + c(x)y = f(x),$$

and if $c_1 y_1(x) + c_2 y_2(x)$ is the general solution of the associated homogeneous differential equation

$$(3.3) \qquad a(x)y'' + b(x)y' + c(x)y = 0,$$

the general solution of (3.2) *is*

$$(3.4) \qquad\qquad y_0(x) + c_1 y_1(x) + c_2 y_2(x).$$

To prove the theorem, let $y(x)$ be an arbitrary solution of equation (3.2). We have then the two identities:

$$a(x)y_0''(x) + b(x)y_0'(x) + c(x)y_0(x) \equiv f(x),$$
$$a(x)y''(x) + b(x)y'(x) + c(x)y(x) \equiv f(x).$$

Subtracting the first of these from the second yields

$$a(x)[y(x) - y_0(x)]'' + b(x)[y(x) - y_0(x)]' + c(x)[y(x) - y_0(x)] \equiv 0;$$

that is, the difference $y(x) - y_0(x)$ is a solution of the homogeneous equation (3.3). Thus,

$$y(x) - y_0(x) = c_1 y_1(x) + c_2 y_2(x),$$

for some choice of the constants c_1 and c_2, and (3.4) follows at once. Conversely, it is easy to verify that (3.4) is a solution of (3.2).

As an example, consider the differential equation

$$(3.5) \qquad\qquad y'' - 3y' + 2y = 4.$$

Its associated homogeneous equation is

$$(3.6) \qquad\qquad y'' - 3y' + 2y = 0,$$

the general solution of which is

$$c_1 e^x + c_2 e^{2x}.$$

By inspection, we observe that a particular solution $y_0(x)$ of (3.5) is the function 2. It follows at once that the general solution of (3.5) may be written as

$$2 + c_1 e^x + c_2 e^{2x}.$$

When a particular solution of an inhomogeneous differential equation cannot be found by inspection, other methods are available. A general method, called *variation of parameters*, is available. But we shall begin with certain special devices that are frequently useful. Roughly speaking, we can apply these devices when the right-hand member of the equation is composed of the sum of terms each of which possesses a finite number of essentially different derivatives. For example, to solve a differential equation like

$$(3.7) \qquad\qquad y'' - y' - 2y = -2e^x - 10 \cos x$$

we first write down the corresponding homogeneous differential equation and its general solution:

$$y'' - y' - 2y = 0,$$
$$c_1 e^{-x} + c_2 e^{2x}.$$

Next, we note that the right-hand member of (3.7) contains the terms e^x and $\cos x$ (except for constant multipliers). Except for constant factors, these terms and their possible derivatives are e^x, $\sin x$, $\cos x$. The plan is to try to determine constants a, b, and c so that the function

$$ae^x + b \sin x + c \cos x$$

is a solution of (3.7). If this function is substituted for y in (3.7), we have

$$-2ae^x + (c - 3b) \sin x - (3c + b) \cos x = -2e^x - 10 \cos x.$$

This equation will be an identity if

$$-2a = -2,$$
$$c - 3b = 0,$$
$$3c + b = 10.$$

It follows that $a = 1$, $b = 1$, $c = 3$, and one solution of (3.7) is

$$e^x + \sin x + 3 \cos x.$$

Since the general solution of the corresponding homogeneous equation

$$y'' - y' - 2y = 0$$

is

$$c_1 e^{-x} + c_2 e^{2x},$$

the general solution of (3.7) is

$$e^x + \sin x + 3 \cos x + c_1 e^{-x} + c_2 e^{2x}.$$

Here is another example. Solve the differential equation

(3.8) $$y'' + 4y = -4x^2 + 2.$$

The general solution of the corresponding homogeneous equation is seen to be $c_1 \sin 2x + c_2 \cos 2x$. We shall try for a solution of (3.8) of the form $y = ax^2 + bx + c$. We have

$$4ax^2 + 4bx + (2a + 4c) = -4x^2 + 2.$$

Thus, we try to determine constants a, b, and c such that

$$4a = -4,$$
$$4b = 0,$$
$$2a + 4c = 2.$$

It follows that we may set $a = -1$, $b = 0$, $c = 1$, and a particular solution of (3.8) is then $-x^2 + 1$. Its general solution is

$$-x^2 + 1 + c_1 \sin 2x + c_2 \cos 2x.$$

Next, consider the differential equation

(3.9) $$y'' - y = 2e^x.$$

The general solution of the corresponding homogeneous equation

(3.10) $$y'' - y = 0$$

is

$$c_1 e^x + c_2 e^{-x}.$$

We may try to determine the constant $a \neq 0$ so that $y = ae^x$ is a solution of (3.9). Upon substitution in (3.9) we have the equation

$$0 = 2e^x$$

for the determination of the constant a. This is not possible, and another method of solution must be attempted.

The method we have been employing failed for equation (3.9) because the term e^x in the right-hand member is also a solution of the homogeneous equation (3.10). The difficulty can be circumvented by the device of determining a constant $a \neq 0$ such that $y = axe^x$ is a solution of (3.9). We have then

$$2ae^x = 2e^x.$$

It follows that $a = 1$ and that a particular solution of (3.9) is xe^x. Its general solution is, then,

$$xe^x + c_1 e^x + c_2 e^{-x}.$$

As a final example of this kind, consider the differential equation

(3.11) $$y'' + y = 2e^x + 4 \sin x.$$

The general solution of the corresponding homogeneous equation is $c_1 \sin x + c_2 \cos x$. We observe that the term $4 \sin x$ in the right-hand member of (3.11) is a solution of the corresponding homogeneous equation.

This suggests that we try for a particular solution of (3.11) of the form

$$(3.12) \qquad y = ae^x + bx \sin x + cx \cos x,$$

where a, b, and c are constants. Upon substituting (3.12) in (3.11) we have

$$2ae^x + 2b \cos x - 2c \sin x = 2e^x + 4 \sin x,$$

and we may take

$$a = 1, \qquad b = 0, \qquad c = -2.$$

The general solution of (3.11) is, accordingly,

$$e^x - 2x \cos x + c_1 \sin x + c_2 \cos x.$$

The method above would fail, for example, if the equation were

$$y'' + y = \tan x,$$

since $\tan x$ does not have only a finite number of essentially different forms for its derivatives.

Exercises

Find the general solutions of the given differential equations.

1. $y'' - 3y' + 2y = \sin x.$

2. $y'' + 2y' + 2y = 1 + x^2.$

3. $y'' + 4y' + 4y = x - 2e^{2x}.$

4. $y'' + 4y' + 8y = x - e^x.$

5. $y'' - 6y' + 9y = e^x \sin x.$

6. $y'' - 4y' + 3y = x^3.$

7. $8y'' + 4y' + y = \sin x - 2 \cos x.$

8. $4y'' + 4y' + y = e^x - 2 \cos 2x.$

Find the general solution of the differential equations in Exercises 9–12. Then find the solution determined by the given conditions.

9. $y'' + y = 3 \cos x, \; y(0) = 0, \; y'(0) = 1.$

10. $y'' + y = e^x + 3 \cos x, \; y(0) = \frac{3}{2}, \; y'(0) = -\frac{3}{2}.$

11. $y'' - 3y' + 2y = 2 + e^x, \; y(0) = 1, \; y'(0) = 3.$

12. $y'' - 9y = 18x + 6e^{3x}, \; y(0) = -1, \; y'(0) = -1.$

Answers

1. $\frac{1}{10} \sin x + \frac{3}{10} \cos x + c_1 e^x + c_2 e^{2x}$.

2. $1 - x + \frac{1}{2}x^2 + e^{-x}(c_1 \cos x + c_2 \sin x)$.

3. $-\frac{1}{4} + \frac{1}{4}x - \frac{1}{2}e^{2x} + c_1 e^{-2x} + c_2 x e^{-2x}$.

4. $-\frac{1}{16} + \frac{1}{8}x - \frac{1}{13}e^x + e^{-2x}(c_1 \cos 2x + c_2 \sin 2x)$.

5. $e^x(\frac{3}{25} \sin x + \frac{4}{25} \cos x) + c_1 e^{3x} + c_2 x e^{3x}$.

6. $\frac{80}{27} + \frac{26}{9}x + \frac{4}{3}x^2 + \frac{1}{3}x^3 + c_1 e^x + c_2 e^{3x}$.

7. $\frac{2}{13} \cos x - \frac{3}{13} \sin x + e^{-x/4}(c_1 \cos \frac{1}{4}x + c_2 \sin \frac{1}{4}x)$.

8. $\frac{1}{9}e^x + \frac{30}{289} \cos 2x - \frac{16}{289} \sin 2x + c_1 e^{-x/2} + c_2 x e^{-x/2}$.

9. $\frac{3}{2}x \sin x + c_1 \cos x + c_2 \sin x; \frac{3}{2}x \sin x + \sin x$.

10. $\frac{1}{2}e^x + \frac{3}{2}x \sin x + c_1 \cos x + c_2 \sin x; \frac{1}{2}e^x + \frac{3}{2}x \sin x + \cos x - 2 \sin x$.

11. $1 - xe^x + c_1 e^x + c_2 e^{2x}; 1 - xe^x - 4e^x + 4e^{2x}$.

12. $xe^{3x} - 2x + c_1 e^{3x} + c_2 e^{-3x}; xe^{3x} - 2x - \frac{1}{2}e^{3x} - \frac{1}{2}e^{-3x}$.

4   Variation of parameters

When the general solution of the homogeneous equation

$$(4.1) \qquad a(x)y'' + b(x)y' + c(x)y = 0$$

is known, the method of *variation of parameters* always provides a particular solution $y_0(x)$ of the inhomogeneous equation

$$(4.2) \qquad a(x)y'' + b(x)y' + c(x)y = f(x).$$

As usual, we are assuming that $a(x)$, $b(x)$, $c(x)$, and $f(x)$ are continuous, with $a(x) \neq 0$, on an interval I. Recall that if $c_1 y_1(x) + c_2 y_2(x)$ is the general solution of (4.1) and $y_0(x)$ is a particular solution of (4.2), the general solution of (4.2) is given by

$$(4.3) \qquad y_0(x) + c_1 y_1(x) + c_2 y_2(x).$$

To obtain such a particular solution $y_0(x)$, suppose that $y_1(x)$ and $y_2(x)$ are linearly independent solutions of equation (4.1). We shall show that functions $u_1(x)$ and $u_2(x)$ can be determined so that

$$y_0(x) = u_1(x)y_1(x) + u_2(x)y_2(x)$$

is a particular solution of (4.2) and such that

$$(4.4) \qquad u_1'(x)y_1(x) + u_2'(x)y_2(x) = 0.$$

[It is the auxiliary condition (4.4) that leads to the name "variation of parameters."] To that end substitute $y_0(x)$ in (4.2), and, using (4.4), obtain

$$(4.5) \qquad u_1'(x)y_1'(x) + u_2'(x)y_2'(x) = \frac{f(x)}{a_0(x)}.$$

Equations (4.4) and (4.5) may be solved for $u_1'(x)$ and $u_2'(x)$:

$$(4.6) \qquad u_1'(x) = -\frac{y_2(x)f(x)}{a_0(x)w(x)}, \qquad u_2'(x) = \frac{y_1(x)f(x)}{a_0(x)w(x)},$$

where $w(x) = y_1(x)y_2'(x) - y_1'(x)y_2(x)$ is the wronskian of $y_1(x)$ and $y_2(x)$. From equations (4.6) we have

$$u_1(x) = -\int_{x_0}^{x} \frac{y_2(t)f(t)}{a_0(t)w(t)}\, dt, \qquad u_2(x) = \int_{x_0}^{x} \frac{y_1(t)f(t)}{a_0(t)w(t)}\, dt.$$

The point $x = x_0$ may be any convenient point of I. A particular solution of (4.2) is then

$$y_0(x) = u_1(x)y_1(x) + u_2(x)y_2(x).$$

This may be written in the symmetric form

$$(4.7) \qquad y_0(x) = \int_{x_0}^{x} \begin{vmatrix} y_1(t) & y_2(t) \\ y_1(x) & y_2(x) \end{vmatrix} \frac{f(t)}{a_0(t)w(t)}\, dt.$$

Recall that $w(t)$ is not equal to zero on I.

   Example.   Using variation of parameters, find the general solution of

$$(4.8) \qquad y'' + y = x.$$

Here, the general solution of the associated homogeneous differential equation is $c_1 \cos x + c_2 \sin x$, $f(x) = x$, $a_0(x) = 1$. If we set

$$y_1(x) = \cos x, \qquad y_2(x) = \sin x,$$

the wronskian $w(x) = 1$. It is often simpler to employ equations (4.4) and (4.5) at this point rather than (4.7). Those equations become

$$(4.4)' \qquad u_1'(x) \cos x + u_2'(x) \sin x = 0,$$

$$(4.5)' \qquad u_1'(x)(-\sin x) + u_2'(x) \cos x = x.$$

Solving these equations for $u_1'(x)$ and $u_2'(x)$ we obtain

$$u_1'(x) = -x \sin x, \qquad u_2'(x) = x \cos x,$$

and, thus,

$$u_1(x) = \int -x \sin x \, dx = x \cos x - \sin x,$$

$$u_2(x) = \int x \cos x \, dx = x \sin x + \cos x.$$

A particular solution $y_0(x)$ is then

$$y_0(x) = u_1(x)y_1(x) + u_2(x)y_2(x)$$
$$= x.$$

The general solution of (4.8) may then be written

$$x + c_1 \cos x + c_2 \sin x.$$

In this example, if we take $x_0 = 0$, $y_1(x) = \cos x$, $y_2(x) = \sin x$, $w(x) = 1$, (4.7) becomes

$$y_0(x) = \int_0^x \begin{vmatrix} \cos t & \sin t \\ \cos x & \sin x \end{vmatrix} t \, dt$$

$$= \int_0^x t \sin (x - t) \, dt$$

$$= x - \sin x.$$

This choice $y_0(x)$ yields the same general solution as before. In this instance, we could, of course, have guessed the particular solution x by inspection of (4.8).

The following example is more substantial.

Example. To find the general solution of the differential equation

$$(4.9) \qquad\qquad y'' + y = \csc x \qquad (0 < x < \pi)$$

we note first that the general solution of the corresponding homogeneous equation

$$(4.10) \qquad\qquad y'' + y = 0$$

is $c_1 \cos x + c_2 \sin x$. We take $y_1(x) = \cos x$, $y_2(x) = \sin x$, $x_0 = \pi/2$. The wronskian $w(t)$ of (4.7) is

$$\begin{vmatrix} \cos t & \sin t \\ -\sin t & \cos t \end{vmatrix} = 1$$

so that (4.7) becomes

$$y_0(x) = \int_{\pi/2}^x \begin{vmatrix} \cos t & \sin t \\ \cos x & \sin x \end{vmatrix} \csc t \, dt.$$

That is,

$$y_0(x) = \sin x \int_{\pi/2}^x \frac{\cos t}{\sin t} dt - \cos x \int_{\pi/2}^x dt,$$

or

(4.11) $$y_0(x) = \sin x \ln (\sin x) + \left(\frac{\pi}{2} - x\right) \cos x.$$

The general solution of (4.9) is then

$$\sin x \ln (\sin x) + \left(\frac{\pi}{2} - x\right) \cos x + c_1 \cos x + c_2 \sin x.$$

or

$$\sin x \ln (\sin x) - x \cos x + c_1 \cos x + c_2 \sin x.$$

Equation (4.9) cannot be solved by the methods of the last section.

Exercises

1. Find the general solution of $y'' + y = 2 - x$ both by guessing a particular solution and by finding a particular solution by variation of parameters.

2. Find a solution as in Exercise 1 for $y'' - 3y' + 2y = 2$.

3. Find a solution as in Exercise 1 for $y'' + 4y = x$.

4. Find a solution as in Exercise 1 for $y'' + 4y = e^x$.

    Using variation of parameters find the general solution of the given equations.

5. $y'' + y = \tan x$.

6. $y'' + y = \sec x$.

7. $y'' + y = 2 \sin x$.

8. $y'' + 4y = 4x + 6$.

9. $y'' - 2y' - 3y = 9x - 3$.

10. $y'' - y = \sin x$.

11. $y'' + 4y' + 4y = -8.$

12. $y'' - y = 1 - x.$

13. $y'' - y = 2e^x.$

14. $y'' - 2y' + 2y = 2x + 4.$

15. $y'' + y = \cot x.$

16.   Substitute the function (4.11) in (4.9) and verify that it is a solution.

Answers

1. $2 - x + c_1 \sin x + c_2 \cos x.$

2. $1 + c_1 e^x + c_2 e^{2x}.$

3. $\dfrac{x}{4} + c_1 \sin 2x + c_2 \cos 2x.$

4. $\frac{1}{5} e^x + c_1 \sin 2x + c_2 \cos 2x.$

5. $c_1 \sin x + c_2 \cos x - \cos x \ln |\sec x + \tan x|.$

6. $c_1 \sin x + c_2 \cos x + x \sin x + \cos x \ln |\cos x|.$

7. $c_1 \sin x + c_2 \cos x - x \cos x.$

9. $-3x + 3 + c_1 e^{-x} + c_2 e^{3x}.$

11. $2 + e^{-2x}(c_1 + c_2 x).$

13. $xe^x + c_1 e^x + c_2 e^{-x}.$

15. $\sin x \ln |\csc x - \cot x| + c_1 \cos x + c_2 \sin x.$

5   Euler-type equations

In the second-order case, Euler-type differential equations are equations of the form

(5.1) $ax^2 y'' + bxy' + cy = f(x) \qquad (a \neq 0),$

where a, b, and c are constants, and $f(x)$ is continuous on an interval I. We shall first examine the homogeneous equation

(5.2) $ax^2 y'' + bxy' + cy = 0 \qquad (a \neq 0).$

Note that when equation (5.2) is solved for y'' we have

$$y'' = -\frac{b}{ax} y' - \frac{c}{ax^2} y;$$

accordingly, the fundamental existence theorem applies to intervals of the x-axis that do not contain the origin.

The substitution $x = e^t$ (when $x > 0$), or $x = -e^t$ (when $x < 0$), reduces an equation (5.2) [or (5.1)] to a linear differential equation with constant coefficients.

An example will clarify the method.

Example.   Find the general solution of the differential equation

(5.3) $$x^2 y'' - 2xy' + 2y = 0 \qquad (x > 0).$$

Under the transformation $x = e^t$ we have

$$y' = \frac{dy}{dx} = \frac{dy}{dt} \Big/ \frac{dx}{dt} = \frac{1}{x} \frac{dy}{dt},$$

(5.4)

$$y'' = \frac{d}{dx}\left(\frac{dy}{dx}\right) = \frac{d}{dt}\left(\frac{dy}{dx}\right) \Big/ \frac{dx}{dt} = \frac{1}{x^2}\left(\frac{d^2 y}{dt^2} - \frac{dy}{dt}\right).$$

These equations then become†

$$xy' = \dot{y}, \qquad x^2 y'' = \ddot{y} - \dot{y}.$$

Equation (5.3) becomes

$$\frac{d^2 y}{dt^2} - 3\frac{dy}{dt} + 2y = 0.$$

Its general solution is

$$y = c_1 e^t + c_2 e^{2t}.$$

Thus, the general solution of (5.3) is,

(5.5) $$y = c_1 x + c_2 x^2.$$

Example.   Solve the differential equation

(5.6) $$x^2 y'' + xy' + y = \ln x \qquad (x > 0).$$

---

† We shall follow the custom of using primes to designate differentiation with respect to x and dots to indicate differentiation with respect to t.

We transform (5.6) by means of the substitution $x = e^t$. Using equations (5.4) we have

(5.7) $$\frac{d^2y}{dt^2} + y = t.$$

The corresponding homogeneous equation has the general solution

$$c_1 \sin t + c_2 \cos t.$$

We next try for a particular solution of (5.7) by attempting to determine constants a and b so that $at + b$ is a solution. A substitution in (5.7) leads to the conditions $a = 1$, $b = 0$, so that t is seen to be the particular solution we seek (this is clear by inspection). Accordingly, the general solution of (5.7) is

$$t + c_1 \sin t + c_2 \cos t,$$

and, consequently, the general solution of (5.6) is

$$\ln x + c_1 \sin \ln x + c_2 \cos \ln x \qquad (x > 0).$$

If the right-hand member of (5.6) had been $\ln(-x)$ and we were interested in solutions for $x < 0$, the only change in the answer would be to replace x by $-x$ or by $|x|$ throughout.

Exercises

Find the general solution of the following differential equations.

1. $x^2y'' - 3xy' + 3y = 0.$

2. $x^2y'' - xy' + 2y = 0.$

3. $x^2y'' + 3xy' + 2y = 0.$

4. $x^2y'' - 2y = 0.$

5. $x^2y'' + 5xy' + 8y = 0.$

6. $x^2y'' + xy' - 4y = 0.$

7. $x^2y'' - xy' - 3y = 0.$

8. $x^2y'' - 5xy' + 13y = 0.$

9. $x^2y'' + xy' + 9y = 0.$

10. $x^2y'' - 3xy' + 3y = \ln|x|.$

11. $x^2y'' - xy' + 2y = 1 + \ln^2|x|.$

12. $x^2y'' + xy' + y = 3 \sin (\ln x^2)$ $(x > 0)$.

13. $x^2y'' - xy' + 2y = 2x^2 - 3x + 2$ $(x < 0)$.

14. $x^2y'' + xy' - 9y = -16x$.

15.* $x^2y'' + xy' + y = \csc (\ln x)$ $(1 < x < e^{\pi})$.

Answers

1. $c_1x + c_2x^3$.

3. $x^{-1}(c_1 \cos \ln |x| + c_2 \sin \ln |x|)$.

5. $x^{-2}(c_1 \cos \ln x^2 + c_2 \sin \ln x^2)$.

7. $c_1x^3 + c_2x^{-1}$.

9. $c_1 \cos (3 \ln |x|) + c_2 \sin (3 \ln |x|)$.

10. $\frac{4}{9} + \frac{1}{3} \ln |x| + c_1x + c_2x^3$.

11. $1 + \ln |x| + \frac{1}{2} \ln^2 |x| + x(c_1 \cos \ln |x| + c_2 \sin \ln |x|)$.

13. $x^2 - 3x + 1 + x[c_1 \cos \ln (-x) + c_2 \sin \ln (-x)]$.

15. $[\sin \ln x][\ln (\sin \ln x)] - (\ln x) \cos \ln x + c_1 \cos \ln x + c_2 \sin \ln x$.

6 The wronskian

In this section we shall prove the basic theorems we have been employing earlier in this chapter. Recall that the wronskian of two solutions $u(x)$ and $v(x)$ of a differential equation

$$(6.1) \qquad y'' + b(x)y' + c(x)y = 0$$

has been defined as the determinant†

$$(6.2) \qquad w(x) = \begin{vmatrix} u(x) & v(x) \\ u'(x) & v'(x) \end{vmatrix} = u(x)v'(x) - v(x)u'(x).$$

As usual, we assume that $b(x)$ and $c(x)$ are continuous functions on an interval I.

We shall begin by borrowing a lemma from linear algebra.

---

† The wronskian is also the determinant

$$\begin{vmatrix} v & u \\ v' & u' \end{vmatrix} = vu' - v'u.$$

This seeming ambiguity will not be troublesome as long as it is clear which of the two is being used at a given moment.

Lemma 6.1. *A necessary and sufficient condition that there exist a pair of numbers* c_1 *and* c_2, *not both zero, satisfying the simultaneous equations*

$$(6.4) \qquad\qquad c_1 a + c_2 b = 0,$$

$$(6.5) \qquad\qquad c_1 h + c_2 k = 0,$$

is that the determinant of the coefficients of c_1 *and* c_2

$$(6.6) \qquad\qquad \begin{vmatrix} a & b \\ h & k \end{vmatrix} = 0.$$

We continue with the following theorem.

Theorem 6.1. *If* $u(x)$ *and* $v(x)$ *are differentiable functions that are linearly dependent on an interval* I, *their wronskian is identically zero on* I.

Note that we have not assumed that $u(x)$ and $v(x)$ are solutions of a differential equation. Theorem 6.1 may be proved as follows. By hypothesis, there exist constants c_1 and c_2, not both zero, such that

$$(6.7) \qquad\qquad c_1 u(x) + c_2 v(x) \equiv 0 \qquad (\text{on } I).$$

It follows that

$$(6.7)' \qquad\qquad c_1 u'(x) + c_2 v'(x) \equiv 0 \qquad (\text{on } I).$$

Let $x = x_0$ be an arbitrary point of I. Then (6.7) and (6.7)' hold at that point:

$$(6.8) \qquad\qquad \begin{aligned} c_1 u(x_0) + c_2 v(x_0) &= 0, \\ c_1 u'(x_0) + c_2 v'(x_0) &= 0. \end{aligned}$$

Thus, there exist constants c_1 and c_2, not both zero, such that equations (6.8) hold. It follows from Lemma 6.1 that the determinant

$$\begin{vmatrix} u(x_0) & v(x_0) \\ u'(x_0) & v'(x_0) \end{vmatrix} = 0;$$

that is, the wronskian of $u(x)$ and $v(x)$ vanishes at $x = x_0$. But $x = x_0$ was an arbitrary point of I; accordingly, the wronskian is zero at every point of I, and the theorem is proved.

Corollary. *If two solutions of equation* (6.1) *are linearly dependent on* I, *their wronskian vanishes identically on* I.

The converse of Theorem 6.1 is not valid; however, we shall next prove a converse of the corollary.

Theorem 6.2. If the wronskian of two solutions $u(x)$ and $v(x)$ of equation (6.1) is identically zero on I, the solutions are linearly dependent.

We have given solutions $u(x)$ and $v(x)$ of equation (6.1) such that

$$\begin{vmatrix} u(x) & v(x) \\ u'(x) & v'(x) \end{vmatrix} \equiv 0 \qquad \text{(on } I).$$

Let $x = x_0$ be any point of I, and it follows that

$$\begin{vmatrix} u(x_0) & v(x_0) \\ u'(x_0) & v'(x_0) \end{vmatrix} = 0.$$

Lemma 6.1 then asserts that there exist constants c_1 and c_2, not both zero, such that the equations

$$(6.8) \qquad \begin{aligned} c_1 u(x_0) + c_2 v(x_0) &= 0, \\ c_1 u'(x_0) + c_2 v'(x_0) &= 0 \end{aligned}$$

are satisfied. Consider the solution

$$c_1 u(x) + c_2 v(x),$$

where c_1 and c_2 are such a pair of constants. According to (6.8), this solution and its derivative vanish at $x = x_0$. By the corollary to the fundamental existence theorem, such a solution is identically zero.

We have thus shown that there exist constants c_1 and c_2, not both zero, such that

$$c_1 u(x) + c_2 v(x) \equiv 0 \qquad \text{(on } I);$$

that is, $u(x)$ and $v(x)$ are linearly dependent on I.

The proof of Theorem 6.2 is complete. We may now combine Theorem 6.2 and the corollary to Theorem 6.1 into the following theorem.

Theorem 6.3. A necessary and sufficient condition that two solutions of the differential equation (6.1) be linearly dependent on I is that their wronskian vanish identically on I.

An example will demonstrate that the converse of Theorem 6.1 is not valid. To that end consider the pair of functions x^2 and $x\sqrt{x^2}$ on the interval $I: -1 \le x \le 1$. Their wronskian is

$$\begin{vmatrix} x^2 & x\sqrt{x^2} \\ 2x & 2\sqrt{x^2} \end{vmatrix} \equiv 0.$$

Now suppose that constants c_1 and c_2 exist such that

$$c_1 x^2 + c_2 x\sqrt{x^2} \equiv 0 \qquad \text{(on } I).$$

This identity must then hold when $x = 1$ and when $x = -1$. That is, we must have

$$c_1 + c_2 = 0,$$
$$c_1 - c_2 = 0.$$

It follows that $c_1 = c_2 = 0$, and, hence, that the given functions are linearly independent on I.

Theorem 6.4.   The wronskian of two solutions of the differential equation

(6.1) $$y'' + b(x)y' + c(x)y = 0$$

is either identically zero, or never zero on I.

We are assuming, as usual, that $b(x)$ and $c(x)$ are continuous functions on I. Before beginning the proof, recall (Chapter 2, Section 3, Corollary 2) that, if a solution of the first-order linear differential equation

(6.9) $$w' + b(x)w = 0$$

vanishes at a single point of I, then it is identically zero on I.

Consider then the wronskian of two solutions $u(x)$ and $v(x)$ of equation (6.1),

$$w(x) = \begin{vmatrix} u(x) & v(x) \\ u'(x) & v'(x) \end{vmatrix} = u(x)v'(x) - v(x)u'(x).$$

Then,

$$\begin{aligned} w'(x) &= u(x)v''(x) - v(x)u''(x) \\ &= u(x)[-b(x)v'(x) - c(x)v(x)] - v(x)[-b(x)u'(x) - c(x)u(x)] \\ &= -b(x)[u(x)v'(x) - v(x)u'(x)] \\ &= -b(x)w(x). \end{aligned}$$

That is, $w(x)$ is a solution of the differential equation (6.9). It follows at once that the wronskian is either identically zero, or never zero.

7   Linear equations of higher order

The theorems given in the preceding sections of this chapter all have natural generalizations to linear differential equations of order n. We shall state these extensions without proof.

We are now considering differential equations

(7.1) $$a_0(x)y^{(n)} + a_1(x)y^{(n-1)} + \cdots + a_{n-1}(x)y' + a_n(x)y = f(x),$$

where $a_0(x)$, $a_1(x), \ldots, a_n(x)$, and $f(x)$ are continuous, with $a_0(x) \neq 0$, on an interval I of the x-axis, and n is a positive integer. When $f(x) \equiv 0$, we have the homogeneous differential equation

$$(7.2) \qquad a_0(x)y^{(n)} + a_1(x)y^{(n-1)} + \cdots + a_{n-1}(x)y' + a_n(x)y = 0.$$

Linear dependence is now defined as follows: n solutions $y_1(x)$, $y_2(x), \ldots,$ $y_n(x)$ are said to be linearly dependent if there exist constants $c_1, c_2, \ldots, c_n$, not all zero, such that

$$(7.3) \qquad c_1 y_1(x) + c_2 y_2(x) + \cdots + c_n y_n(x) \equiv 0$$

on I. If, however, the identity (7.3) implies that $c_1 = c_2 = \cdots = c_n = 0$, the n solutions are said to be linearly independent.

The *wronskian* of n solutions $y_1(x)$, $y_2(x), \ldots, y_n(x)$ is the determinant

$$(7.4) \qquad \begin{vmatrix} y_1(x) & y_2(x) & \cdots & y_n(x) \\ y_1'(x) & y_2'(x) & \cdots & y_n'(x) \\ \cdots & \cdots & \cdots & \cdots \\ y_1^{(n-1)}(x) & y_2^{(n-1)}(x) & \cdots & y_n^{(n-1)}(x) \end{vmatrix}.$$

Theorem 7.1. *The wronskian of n solutions of equation (7.2) is either identically zero, or is never zero, on I.*

Theorem 7.2. *A necessary and sufficient condition that n solutions of equation (7.2) be linearly dependent is that their wronskian vanish identically on I.*

Theorem 7.3. *There exist infinitely many sets of n linearly independent solutions of equation (7.2).*

Theorem 7.4. *If $y_1(x)$, $y_2(x), \ldots, y_n(x)$ are n linearly independent solutions of equation (7.2), the general solution of (7.2) can be written in the form*

$$c_1 y_1(x) + c_2 y_2(x) + \cdots + c_n y_n(x),$$

where $c_1, c_2, \ldots, c_n$ are constants.

Theorem 7.5. *If $y_0(x)$ is any solution of equation (7.1), the general solution of (7.1) is given by*

$$y_0(x) + c_1 y_1(x) + c_2 y_2(x) + \cdots + c_n y_n(x),$$

where $c_1 y_1(x) + c_2 y_2(x) + \cdots + c_n y_n(x)$ is the general solution of equation (7.2).

The proofs of all these theorems are strictly analogous to the proofs of the corresponding theorems for the case $n = 2$.

Constant coefficients. The homogeneous differential equation

$$(7.5) \qquad a_0 y^{(n)} + a_1 y^{(n-1)} + \cdots + a_{n-1} y' + a_n y = 0 \qquad (a_0 \neq 0),$$

where $a_0, a_1, \ldots, a_n$ are constants, is solved in a manner similar to that for the case $n = 2$. We try for a solution of the form

$$y = e^{mx}.$$

Substituting this and its derivatives in equation (7.5) now leads to the *indicial equation*

$$(7.6) \qquad a_0 m^n + a_1 m^{n-1} + \cdots + a_{n-1} m + a_n = 0,$$

a polynomial equation of degree n in m. Such an equation has precisely n roots m, when multiple roots are counted according to their multiplicity. The problem of finding these n roots is, in general, not easy.

An equation such as

$$(7.5)' \qquad y''' - 6y'' + 11y' - 6y = 0$$

can, however, be solved readily. The corresponding indicial equation is

$$(7.6)' \qquad m^3 - 6m^2 + 11m - 6 = 0$$

the roots of which are $m = 1, 2, 3$. Thus, solutions of $(7.5)'$ are

$$e^x, \qquad e^{2x}, \qquad e^{3x}.$$

The wronskian of these solutions is the determinant

$$\begin{vmatrix} e^x & e^{2x} & e^{3x} \\ e^x & 2e^{2x} & 3e^{3x} \\ e^x & 4e^{2x} & 9e^{3x} \end{vmatrix} = 2e^{6x}.$$

The solutions are then linearly independent, and the general solution of $(7.5)'$ is

$$c_1 e^x + c_2 e^{2x} + c_3 e^{3x}.$$

The equation

$$(7.7) \qquad y''' - 6y'' + 11y' - 6y = 12$$

is readily seen to have the particular solution $y = -2$. Its general solution, by Theorem 7.5, is then

$$-2 + c_1 e^x + c_2 e^{2x} + c_3 e^{3x}.$$

Recall that if $\alpha + i\beta$ (α, β real, $\beta \neq 0$) is a root of equation (7.6), so is $\alpha - i\beta$; that is, imaginary roots occur in conjugate pairs. When the order n of the differential equation (7.5) is greater than or equal to 4, not only real roots, but imaginary roots of (7.6) as well, may be repeated. The method of obtaining linearly independent solutions in these instances will be clear from the following illustration.

Suppose, for example, an equation of the form (7.5) leads to an indicial equation (7.6), the roots of which are

$$1, \quad -1, \quad -1, \quad 2, \quad 2, \quad 2, \quad 2 + 3i, \quad 2 - 3i, \quad 3 + 4i, \quad 3 - 4i, \quad 3 + 4i, \quad 3 - 4i.$$

Here, $n = 12$, and a set of 12 linearly independent solutions of the corresponding differential equation is

$$e^x, \quad e^{-x}, \quad xe^{-x}, \quad e^{2x}, \quad xe^{2x}, \quad x^2 e^{2x}, \quad e^{2x}\cos 3x, \quad e^{2x}\sin 3x,$$
$$e^{3x}\cos 4x, \quad e^{3x}\sin 4x, \quad xe^{3x}\cos 4x, \quad xe^{3x}\sin 4x.$$

If n had been 14, and if there were another pair of roots $3 + 4i$, $3 - 4i$ of the indicial equation, we would have added

$$x^2 e^{3x}\cos 4x, \quad x^2 e^{3x}\sin 4x$$

to the set above.

Finally, an Euler-type equation of order n is of the form

$$(7.7) \qquad a_0 x^n y^{(n)} + a_1 x^{n-1} y^{(n-1)} + \cdots + a_{n-1} xy' + a_n y = f(x) \qquad (a_0 \neq 0),$$

where $a_0, a_1, \ldots, a_n$ are constants, and $f(x)$ is continuous on some interval I. The substitutions

$$x = e^t \qquad or \qquad x = -e^t,$$

according as $x > 0$ or $x < 0$, again reduces equation (7.7) to a linear differential equation with constant coefficients. When either substitution is made, one has

$$(7.8) \qquad xy' = \dot{y}, \qquad x^2 y'' = \ddot{y} - \dot{y}, \qquad x^3 y''' = \dddot{y} - 3\ddot{y} + 2\dot{y},$$
$$x^4 y'''' = \ddddot{y} - 6\dddot{y} + 11\ddot{y} - 6\dot{y}, \qquad \ldots .$$

The student will predict (correctly) that the coefficients in the substitution for $x^5 y^{(5)}$ will be the coefficients obtained by multiplying the polynomials $(k - 4)(k^3 - 6k^2 + 11k - 6)$.

Consider then the Euler-type differential equation

(7.9) $\qquad\qquad x^3 y''' - 3x^2 y'' + 6xy' - 6y = 0 \qquad (x > 0).$

The substitution $x = e^t$ and (7.8) yield the differential equation

$$\dddot{y} - 6\ddot{y} + 11\dot{y} - 6y = 0,$$

linearly independent solutions of which are

$$e^t, \qquad e^{2t}, \qquad e^{3t}.$$

Corresponding linearly independent solutions of (7.9) are then

$$x, \qquad x^2, \qquad x^3,$$

and the general solution of (7.9) is

$$c_1 x + c_2 x^2 + c_3 x^3.$$

The differential equation

(7.10) $\qquad\qquad x^3 y''' - 3x^2 y'' + 6xy' - 6y = 6 \ln x \qquad (x > 0)$

becomes

(7.11) $\qquad\qquad \dddot{y} - 6\ddot{y} + 11\dot{y} - 6y = 6t$

under the transformation $x = e^t$. A particular solution of this last equation is $-t - 11/6$ (try for a solution of the form $\alpha t + \beta$); consequently, the general solution of (7.11) is

$$-t - \frac{11}{6} + c_1 e^t + c_2 e^{2t} + c_3 e^{3t}.$$

The general solution of (7.10) is, accordingly,

$$-\ln x - \frac{11}{6} + c_1 x + c_2 x^2 + c_3 x^3 \qquad (x > 0).$$

Exercises

1. Find the general solutions of the following differential equations:

   (a) $y''' + y' = 0$;
   (b) $y''' - 3y'' + 4y' - 2y = 0$;
   (c) $y''' - 3y' + 2y = 0$;
   (d)* $y'''' - 8y''' + 42y'' - 104y' + 169y = 0$.
      (*Hint.* One root of the indicial equation is $2 + 3i$.)

2.   Find the general solutions of the following differential equations:

(a) $y''' - y'' - y' + y = 0$;
(b)* $y'''' + 4y = 0$;
(c)* $y''''' - 6y''' + 19y'' - 26y' + 18y = 0$.

3.   A homogeneous linear differential equation with constant coefficients has

$$2, \quad 2, \quad 2, \quad 3 - 4i, \quad 3 + 4i, \quad 3 - 4i, \quad 3 + 4i, \quad 3, \quad 3$$

as roots of its auxiliary equation. What is its general solution?

4.   Do the same as Exercise 3 when the roots are

$$2, \quad 2, \quad 3 - 4i, \quad 3 + 4i, \quad 3 - 4i, \quad 3 + 4i, \quad 3 - 4i, \quad 3 + 4i, \quad 7.$$

5.   Note Exercise 1, and find the general solutions of the following differential equations:

(a) $y''' + y' = 3$;
(b) $y''' - 3y'' + 4y' - 2y = \sin x$;
(c) $y''' - 3y' + 2y = x - \cos x$.

6.   Find the general solutions of the following differential equations. (*Hint.* Some of the results in Exercises 1 and 2 above will be helpful.)

(a) $x^3 y''' + 2x^2 y'' - xy' + y = 0$ $(x > 0)$;
(b) $x^3 y''' + 2xy' - 2y = 0$ $(x < 0)$;
(c) $2x^3 y''' + 3x^2 y'' - 4xy' + 2y = 0$ $(x < 0)$.

7.   Note Exercise 6, and find the general solutions of the following differential equations:

(a) $x^3 y''' + 2x^2 y'' - xy' + y = \ln x$ $(x > 0)$;
(b) $x^3 y''' + 2xy' - 2y = 3 - \ln(-x)$ $(x < 0)$;
(c) $2x^3 y''' + 3x^2 y'' - 4xy' + 2y = 3\ln(-x)$ $(x < 0)$.

8.   Evaluate the wronskian of the three functions

$$e^{ax}, \quad e^{bx}, \quad e^{cx},$$

where a, b, and c are constants.

9.*   Consider the third-order linear differential equation

(1) $a_0(x)y''' + a_1(x)y'' + a_2(x)y' + a_3(x)y = f(x)$,

where $a_0(x)$, $a_1(x)$, $a_2(x)$, $a_3(x)$, and $f(x)$ are continuous on an interval I, with $a_0(x) \neq 0$ there. Using variation of parameters, derive the formula

(2) $$y_0(x) = \int_{x_0}^{x} \begin{vmatrix} y_1(t) & y_2(t) & y_3(t) \\ y_1'(t) & y_2'(t) & y_3'(t) \\ y_1(x) & y_2(x) & y_3(x) \end{vmatrix} \frac{f(t)}{a_0(t)w(t)} \, dt$$

for a particular solution of (1).

10.* Let $u(x)$, $v(x)$, and $w(x)$ be solutions of the homogeneous differential equation associated with equation (1) in Exercise 9* and show that their wronskian

$$z(x) = \begin{vmatrix} u(x) & v(x) & w(x) \\ u'(x) & v'(x) & w'(x) \\ u''(x) & v''(x) & w''(x) \end{vmatrix}$$

is a solution of the first-order linear differential equation

$$a_0(x)z' + a_1(x)z = 0$$

and, hence, that the wronskian is either identically zero, or never zero, on I. (*Hint.* Recall the method of differentiating a determinant, or see Chapter 9, Section 2.)

Answers

1. (a) $c_1 + c_2 \cos x + c_3 \sin x$;

(b) $e^x(c_1 + c_2 \cos x + c_3 \sin x)$;

(c) $c_1 e^x + c_2 x e^x + c_3 e^{-2x}$;

(d) $c_1 e^{2x} \cos 3x + c_2 e^{2x} \sin 3x + c_3 x e^{2x} \cos 3x + c_4 x e^{2x} \sin 3x$.

2. (a) $c_1 e^{-x} + c_2 e^x + c_3 x e^x$;

(b) $e^x(c_1 \cos x + c_2 \sin x) + e^{-x}(c_3 \cos x + c_4 \sin x)$;

(c) $e^x(c_1 \cos x + c_2 \sin x) + e^{2x}[c_3 \cos (x\sqrt{5}) + c_4 \sin (x\sqrt{5})]$.

3. $c_1 e^{2x} + c_2 x e^{2x} + c_3 x^2 e^{2x} + c_4 e^{3x} \cos 4x + c_5 e^{3x} \sin 4x + c_6 x e^{3x} \cos 4x + c_7 x e^{3x} \sin 4x$.

5. (b) $\frac{1}{10} \sin x - \frac{3}{10} \cos x + e^x(c_1 + c_2 \cos x + c_3 \sin x)$.

6. (a) $c_1 x^{-1} + c_2 x + c_3 x \ln x$;

(b) $x[c_1 + c_2 \cos (\ln (-x)) + c_3 \sin (\ln (-x))]$;

(c) $c_1(-x)^{1/2} + c_2 x^{-1} + c_3 x^2$.

7. (b) $-\frac{1}{2} + \frac{1}{2} \ln (-x) + x[c_1 + c_2 \cos \ln (-x) + c_3 \sin \ln (-x)]$.

8. $(b - a)(c - b)(c - a)e^{(a+b+c)x}$.

8 More on guessing particular solutions

The methods for solving inhomogeneous linear differential equations given in Section 3 are very special ones depending on the fact that the terms in the right-hand side of the differential equation have a finite number of essentially different derivative forms. Rules can be set down in such cases to enable the student to make a correct guess as to what to try, but they are cumbersome and of limited value. The differential equation

$$y'' - y' = 2x,$$

for example, requires a slight modification of the procedures given there owing to the absence of the term involving y. Trying for a particular solution of the form

$$y = ax + b$$

will fail, as is readily seen, but the attempt

$$y = ax^2 + bx$$

will succeed.

There is a general method for determining the type of function to try for a particular solution that will always work for the kinds of equations we are presently considering. The method will be clear from the examples below, but first it will be helpful to have the notion of composition of solutions at our disposal.

Composition of solutions.   To find a particular solution of the differential equation

(8.1) $$y'' + 4y = e^x - 2 \sin x + x,$$

we may determine constants c_1, c_2, c_3, c_4, c_5 such that the function

$$c_1 e^x + c_2 \sin x + c_3 \cos x + c_4 x + c_5$$

will be a solution of (8.1). We may also find particular solutions of each of the following differential equations:

$$y'' + 4y = e^x,$$
$$y'' + 4y = -2 \sin x,$$
$$y'' + 4y = x.$$

The sum of these three particular solutions will be seen to be a solution of (8.1). The validity of this method is readily established. The proof for the case of a general second-order linear differential equation is left to the student as an exercise (Exercise 5).

Consider now the differential equation

(8.2) $$y'' + y = 5e^{2x}.$$

The general solution of the corresponding homogeneous equation

(8.2)' $$y'' + y = 0$$

is $c_1 \cos x + c_2 \sin x$. Suppose y is any solution of (8.2). Then, differentiating, we have the identity

$$(8.3) \qquad y''' + y' \equiv 10e^{2x}.$$

Multiplying both members of (8.2) by (-2) and adding to (8.3), we eliminate the right-hand term and we have that y must also be a solution of the homogeneous differential equation

$$(8.4) \qquad y''' - 2y'' + y' - 2y = 0,$$

the characteristic equation of which is

$$(8.5) \qquad m^3 - 2m^2 + m - 2 = 0.$$

Next, note that if y is a solution of the homogeneous equation (8.2)′, it must also be a solution of the equation

$$(8.3)' \qquad y''' + y' = 0.$$

If we multiply both members of (8.2)′ by (-2) and add to (8.3)′, we have that a solution y of the homogeneous equation (8.2)′ is also a solution of (8.4). Equation (8.4), then, has as solutions the general solution of the homogeneous equation (8.2)′ and all particular solutions of the inhomogeneous equation (8.2). It follows that we know two roots of (8.5). They are i and $-i$. Thus, equation (8.5) can be written in factored form as

$$(m^2 + 1)(m - 2) = 0,$$

the third root of which is $m = 2$. Accordingly, the general solution of (8.4) has the form

$$c_1 \cos x + c_2 \sin x + c_3 e^{2x}.$$

The first two terms above provide the general solution of the homogeneous equation (8.2)′, so we try for a particular solution of (8.2) of the form

$$y = c_3 e^{2x}.$$

It is easy to see that $c_3 = 1$, and the general solution of (8.2) is

$$c_1 \cos x + c_2 \sin x + e^{2x}.$$

Consider next the differential equation

$$(8.6) \qquad y'' - 3y' + 2y = 2 \cos x.$$

The general solution of the corresponding homogeneous differential equation is readily seen to be

$$(8.7) \qquad c_1 e^x + c_2 e^{2x}.$$

To eliminate the right-hand member of (8.6) we must differentiate twice, obtaining

$$(8.8) \qquad y^{\text{IV}} - 3y''' + 2y'' = -2 \cos x.$$

Adding (8.6) and (8.8) we have the homogeneous differential equation

$$(8.9) \qquad y^{\text{IV}} - 3y''' + 3y'' - 3y' + 2y = 0,$$

the characteristic equation of which is

$$(8.10) \qquad m^4 - 3m^3 + 3m^2 - 3m + 2 = 0.$$

The argument given in the prior example is valid, and we know that (8.7) is a solution of (8.9). Thus, two roots of equation (8.10) are $1, 2$. It follows that $m^2 - 3m + 2$ is a factor of the left-hand side of (8.10). Using this fact (or, better, synthetic division) we write (8.10) in the form

$$(m^2 - 3m + 2)(m^2 + 1) = 0,$$

the roots of which are $1, 2, i, -i$. The first two roots yield (8.7) so we try for a particular solution of (8.6) of the form

$$c_3 \cos x + c_4 \sin x.$$

The determination of the constants c_3 and c_4 is left to the student.

Finally, consider the differential equation

$$(8.11) \qquad y'' - y' = 2x + 2e^x.$$

The general solution of the corresponding homogeneous differential equation for this example is

$$(8.12) \qquad c_1 e^x + c_2.$$

It will be helpful to use the principle of composition of solutions because the right-hand member of (8.11) contains two different types of functions x and e^x. To that end, consider first the differential equation

$$(8.13) \qquad y'' - y' = 2x.$$

Differentiating (8.13) twice we eliminate the right-hand term and we have

$$(8.14) \qquad y^{\text{IV}} - y''' = 0.$$

The corresponding characteristic equation is

$$m^4 - m^3 = 0,$$

the roots of which are $1, 0, 0, 0$. The general solution of (8.14) is, then, of the form

$$c_1 e^x + c_2 + c_3 x + c_4 x^2.$$

The first two terms yield (8.12) so we try for a particular solution of equation (8.13) of the form

$$c_3 x + c_4 x^2.$$

Substituting this in (8.13) we find that

$$c_3 = -2, \qquad c_4 = -1$$

and a particular solution of (8.13) is

(8.15) $-2x - x^2.$

We turn our attention next to the differential equation

(8.16) $y'' - y' = 2e^x.$

Differentiating once we have

(8.17) $y''' - y'' = 2e^x.$

If we subtract (8.16) from (8.17) we obtain the homogeneous equation

(8.18) $y''' - 2y'' + y' = 0.$

The corresponding characteristic equation is

$$m^3 - 2m^2 + m = 0,$$

or

$$m(m - 1)^2 = 0.$$

Its roots, then, are 0, 1, 1, and the general solution of (8.18) may be written in the form

$$c_1 e^x + c_2 + c_3 x e^x.$$

The first two terms above yield the solution (8.12) of the corresponding homogeneous differential equation so we try for a particular solution of (8.16) of the form

$$c_3 x e^x.$$

Substituting this function in (8.16) we find that $c_3 = 2$. A particular solution of (8.16) is then

(8.19) $2x e^x.$

Using the principle of composition of solutions we may then write the general solution of (8.11) as

$$c_1 e^x + c_2 - 2x - x^2 + 2x e^x.$$

Exercises

Find the general solutions of the given equations.

1. $y'' - 3y' + 2y = e^x - 2e^{2x} + \sin x.$

2. $y'' + y = 4x \sin x + 3 \cos x.$

3. $y'' + 9y = -6 \cos 3x + 20e^x.$

4. $y'' + 2y' + 2y = 2e^x \sin x - \sin x.$

5. Find the general solution of

$$y'' - 2y' + y = 4e^x.$$

6. Establish the validity of the method of composition of solutions for the equation

$$y'' + a(x)y' + b(x)y = f_1(x) + f_2(x).$$

Answers

1. $\frac{1}{10} \sin x + \frac{3}{10} \cos x - xe^x - 2xe^{2x} + c_1 e^x + c_2 e^{2x}.$

3. $2e^x - x \sin 3x + c_1 \cos 3x + c_2 \sin 3x.$

5. $2x^2 e^x + c_1 e^x + c_2 x e^x.$

4

The Laplace transform

Laplace transforms are useful in solving what are known as *initial value problems* when the differential equation, or system, is linear, inhomogeneous, and has constant coefficients. Some examples are

(a)
$$\begin{cases} y'' + y = x, \\ y(0) = 0, \ y'(0) = 1; \end{cases}$$

(b)
$$\begin{cases} \ddot{y} - 3\dot{y} + 2y = \sin t, \\ y(0) = 1, \ y'(0) = 1; \end{cases}$$

(c)
$$\begin{cases} \ddot{x} + 2\dot{x} + x = 2 - e^t, \\ \ddot{y} + 2\dot{y} - y = 1, \\ x(0) = 1, \ \dot{x}(0) = -1, \\ y(0) = 0, \ \dot{y}(0) = 2. \end{cases}$$

These are called initial value problems because the conditions a solution must satisfy are conditions at a single point. Thus,

$$y'' + y = x,$$

$$y(0) = 0, \qquad y'\left(\frac{\pi}{2}\right) = 1$$

is not an initial value problem, because more than one point $\left(x = 0, \ x = \dfrac{\pi}{2}\right)$ is involved in the so-called *side* conditions.

The systems (a) and (b) can be solved by methods with which we are familiar. The method of Laplace transforms will provide an alternate method that is direct and frequently less cumbersome. Systems like (c) will be discussed in Chapter 9.

1   The transform

The Laplace transform $\mathscr{L}[f(x)]$ of a function $f(x)$ is defined by the equation

$$(1.1) \qquad \mathscr{L}[f(x)] = \int_0^\infty e^{-sx} f(x) \, dx = \lim_{t \to \infty} \int_0^t e^{-sx} f(x) \, dx,$$

when this limit exists. Here, s is a sufficiently large constant. In general, the infinite integral in (1.1) will exist if $f(x)$ is a function that does not increase too rapidly as $x \to +\infty$. For example, $\mathscr{L}[f(x)]$ exists when $f(x) = x^\alpha (\alpha > 0)$, but does not exist when $f(x) = e^{x^2}$.

To describe a general class of functions for which the Laplace transform exists we first define what is meant by a *piecewise continuous* function $f(x)$. One such function is a "step function," defined as follows:

$$(1.2) \qquad f(x) = \begin{cases} 1 & (0 \le x < 1), \\ 2 & (1 \le x < 2), \\ \cdots\cdots\cdots\cdots\cdots \\ n & (n - 1 \le x < n) \qquad n = 1, 2, 3, \ldots. \end{cases}$$

Its graph is given in Fig. 4.1.

Now suppose that the positive x-axis can be divided into subintervals on the interior of each of which $f(x)$ is continuous. If on each finite subinterval the limit of $f(x)$ exists as the endpoints are approached from the interior of the subinterval, $f(x)$ is said to be *piecewise continuous* on the positive x-axis.

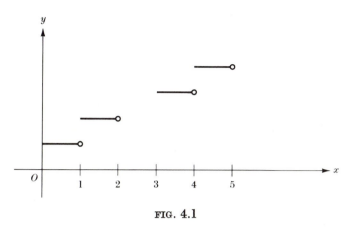

FIG. 4.1

The number of subintervals may be finite or infinite. In Fig. 4.1 there is an infinity of subintervals. The function $f(x)$ defined by the equations

(1.3) $$f(x) = \begin{cases} 2x & (0 \leq x \leq 2) \\ x - 1 & (2 < x < \infty) \end{cases}$$

(see Fig. 4.2) is also piecewise continuous on $[0, \infty)$.

A function that is continuous on $[0, \infty)$ is, of course, also piecewise continuous there.

If $f(x)$ is piecewise continuous on $[0, \infty)$, and if there exist constants M and α such that

$$|f(x)| < Me^{\alpha x} \qquad (0 \leq x < \infty),$$

$f(x)$ is said to be of *exponential order* on $[0, \infty)$. We then have the following theorem from the calculus.

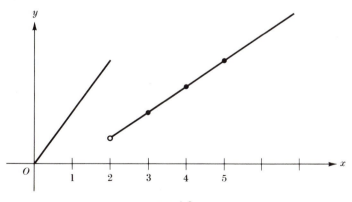

FIG. 4.2

Theorem 1.1.   If $f(x)$ is of exponential order, its Laplace transform exists for $s > \alpha$.

Examples of common functions of exponential order are

(1.4)

1	$\sin ax$	$\sinh ax$	$e^{ax} \sin bx$
x^n	$\cos ax$	$\cosh ax$	$e^{ax} \cos bx$
e^{ax}	$x^n e^{ax}$	$x^n \sinh ax$	$x^n \cosh ax$

where a and b are constants and $n = 0, 1, 2, \ldots$. The functions defined in (1.2) and (1.3) are also of exponential order, and their Laplace transforms exist. Indeed, in the case of the function (1.2), we have

$$\mathcal{L}[f(x)] = \int_0^1 e^{-sx}1 \, dx + \int_1^2 e^{-sx}2 \, dx + \int_2^3 e^{-sx}3 \, dx + \cdots$$

$$= \left[\frac{e^{-sx}}{-s}\right]_0^1 + 2\left[\frac{e^{-sx}}{-s}\right]_1^2 + 3\left[\frac{e^{-sx}}{-s}\right]_2^3 + \cdots$$

$$= \frac{1}{s}[(1 - e^{-s}) + 2(e^{-s} - e^{-2s}) + 3(e^{-2s} - e^{-3s}) + \cdots]$$

$$= \frac{1}{s}[1 + e^{-s} + e^{-2s} + e^{-3s} + \cdots]$$

$$= \frac{1}{s(1 - e^{-s})} \qquad (s > 0).$$

The calculation of the Laplace transform of the function (1.3) is readily accomplished using integration by parts. Indeed, integration by parts is also the principal tool required for the computation of the Laplace transform for the functions (1.4). For example, using integration by parts, we have

$$\mathcal{L}[\sin ax] = \int_0^\infty e^{-sx} \sin ax \, dx$$

$$= \left[-\frac{1}{s}e^{-sx} \sin ax\right]_0^\infty + \frac{a}{s}\int_0^\infty e^{-sx} \cos ax \, dx$$

$$= \frac{a}{s}\int_0^\infty e^{-sx} \cos ax \, dx.$$

Before repeating integration by parts we note from (1.5) that

(1.6) $$\mathcal{L}[\sin ax] = \frac{a}{s}\mathcal{L}[\cos ax].$$

Returning to (1.5) we have

$$\mathcal{L}[\sin ax] = \frac{a}{s}\left\{\left[\frac{e^{-sx} \cos ax}{-s}\right]_0^\infty - \frac{a}{s}\int_0^\infty e^{-sx} \sin ax \, dx\right\}$$

$$= \frac{a}{s}\left\{\frac{1}{s} - \frac{a}{s}\mathcal{L}[\sin ax]\right\}.$$

It follows that

(1.7)
$$\mathscr{L}[\sin ax] = \frac{a}{a^2 + s^2} \qquad (s > 0).$$

From (1.6) we now have also that

(1.8)
$$\mathscr{L}[\cos ax] = \frac{s}{a^2 + s^2} \qquad (s > 0).$$

Table of Laplace Transforms†

$f(x)$	$\mathscr{L}[f(x)]$
1	$\dfrac{1}{s} \quad (s > 0)$
x	$\dfrac{1}{s^2} \quad (s > 0)$
x^n	$\dfrac{n!}{s^{n+1}} \quad (s > 0)$
$\sin ax$	$\dfrac{a}{a^2 + s^2} \quad (s > 0)$
$\cos ax$	$\dfrac{s}{a^2 + s^2} \quad (s > 0)$
e^{ax}	$\dfrac{1}{s - a} \quad (s > a)$
$x^n e^{ax}$	$\dfrac{n!}{(s - a)^{n+1}} \quad (s > a)$
$\sinh ax$	$\dfrac{a}{s^2 - a^2} \quad (s > |a|)$
$\cosh ax$	$\dfrac{s}{s^2 - a^2} \quad (s > |a|)$
$e^{ax} \sin bx$	$\dfrac{b}{(s - a)^2 + b^2} \quad (s > a)$
$e^{ax} \cos bx$	$\dfrac{s - a}{(s - a)^2 + b^2} \quad (s > a)$

(1.9)

(a and b constants, $n = 0, 1, 2, \ldots$)

---

† There are more extensive tables of Laplace transforms to which the student may refer. See, for example, *Standard Mathematical Tables*, 20th ed., Chemical Rubber Publishing Co., Cleveland (1972).

The calculation of $\mathscr{L}[1]$ and $\mathscr{L}[e^{ax}]$ is immediate. To calculate $\mathscr{L}[x^n]$ we employ integration by parts and we have, for $n \geq 1$,

$$\mathscr{L}[x^n] = \int_0^\infty e^{-sx} x^n \, dx = -\frac{1}{s}[x^n e^{-sx}]_0^\infty + \frac{n}{s}\int_0^\infty e^{-sx} x^{n-1} \, dx \qquad (s > 0)$$

$$= \frac{n}{s}\mathscr{L}[x^{n-1}].$$

Applying this formula to $\mathscr{L}[x^{n-1}]$ we have

$$\mathscr{L}[x^n] = \frac{n}{s}\left[\frac{n-1}{s}\mathscr{L}[x^{n-2}]\right].$$

Continuing this process we have, finally, that

$$\mathscr{L}[x^n] = \frac{n!}{s^n}\mathscr{L}[1] = \frac{n!}{s^{n+1}} \qquad (s > 0, \, n = 0, 1, 2, \ldots)$$

(recall that $0! = 1$).

From the last equation we note that

$$\mathscr{L}[x^n] = \int_0^\infty x^n e^{-sx} \, dx = \frac{n!}{s^{n+1}} \qquad (s > 0, \, n = 0, 1, 2, \ldots).$$

Using this result we have that

$$\mathscr{L}[x^n e^{ax}] = \int_0^\infty x^n e^{ax} e^{-sx} \, dx = \int_0^\infty x^n e^{-(s-a)x} \, dx$$

$$= \frac{n!}{(s-a)^{n+1}} \qquad (s > a, \, n = 0, 1, 2, \ldots).$$

The verification of the remaining entries in the table is left to the student in the Exercises.

In our computations we have been using the following results from the calculus (p, s, b, and a constants):

$$\lim_{x \to \infty} x^p e^{-sx} = 0 \qquad (s > 0),$$

$$\lim_{x \to \infty} e^{-sx} \sin bx = 0 \qquad (s > 0),$$

$$\lim_{x \to \infty} e^{-sx} \cos bx = 0 \qquad (s > 0),$$

$$\lim_{x \to \infty} e^{-sx} e^{ax} = 0 \qquad (s > a),$$

$$\lim_{x \to \infty} e^{-sx} x^p \sin bx = 0 \qquad (s > 0),$$

$$\lim_{x \to \infty} e^{-sx} x^p \cos bx = 0 \qquad (s > 0).$$

It should be noted that there are functions that have no Laplace transform. Clearly functions like tan x and sec x are in this category. So is $e^{ax^2}(a > 0)$, for it can be shown that the integral

$$\int_0^\infty e^{-sx} e^{ax^2}\, dx \qquad (a > 0)$$

diverges for every choice of the constant s.

On notation. First, in many applied problems where the use of the Laplace transform is helpful, the independent variable is the time t. It is clear that the Laplace transform of a function $f(t)$ may then be written as

$$\mathscr{L}[f(t)] = \int_0^\infty e^{-st} f(t)\, dt.$$

Traditionally, the derivative $\dot f(t)$ of a function $f(t)$ is indicated by a dot over the f rather than by primes. In recent years the use of primes to indicate differentiation with respect to t has come into rather wide usage. We shall use both notations, when no ambiguity is involved.

An alternate and frequently useful notation for the Laplace transform will be clear from the following examples:

$$F(s) = \int_0^\infty e^{-sx} f(x)\, dx,$$

$$G(s) = \int_0^\infty e^{-sx} g(x)\, dx;$$

that is, capital F indicates the Laplace transform of f, and so on.

In passing we note that

$$F(s) = \int_0^\infty e^{-sx} f(x)\, dx = \int_0^\infty e^{-st} f(t)\, dt = \int_0^\infty e^{-su} f(u)\, du = \int_0^\infty e^{-s*} f(*)\, d*,$$

when the transform of $f(x)$ exists.

Exercises

1. Calculate the Laplace transforms of the following functions using the definition (1.1):

    (a) xe^x, x^2, $x^2 e^x$;

    (b) $e^t \sin t$, $e^t \cos t$.

2. Calculate the Laplace transforms of the following functions using the definition (1.1):

(a) $\sinh ax$;   (b) $\cosh ax$;

(c) $e^{ax} \sin bx$;   (d) $e^{ax} \cos bx$.

3. Calculate the Laplace transforms of the following functions and sketch the curves $y = f(x)$:

(a) $f(x) = \begin{cases} 1 & (0 \le x < 4) \\ 0 & (4 \le x < \infty) \end{cases}$

(b) $f(x) = \begin{cases} 1 & (0 \le x < 1, 2 \le x < 3, 4 \le x < 5, \ldots) \\ 0 & (1 \le x < 2, 3 \le x < 4, 5 \le x < 6, \ldots) \end{cases}$

(c) $f(x) = \begin{cases} x & (0 \le x < 1) \\ x - 1 & (1 \le x < 2) \\ x - 2 & (2 \le x < 3) \\ \cdots\cdots\cdots\cdots\cdots\cdots \\ x - n & (n \le x < n + 1) \\ \cdots\cdots\cdots\cdots\cdots\cdots \end{cases}$

4. Do the same as in Exercise 3 for the following functions:

(a) $f(t) = \begin{cases} 1 & (0 \le t < 1) \\ -1 & (1 \le t < 2) \\ 1 & (2 \le t < 3) \\ -1 & (3 \le t < 4) \\ \cdots\cdots\cdots\cdots\cdots \end{cases}$

(b) $f(t) = |\sin t|$ $(0 \le t < \infty)$.

5. Verify formally that if

$$\mathcal{L}[f] = F(s) = \int_0^\infty e^{-sx} f(x)\, dx,$$

then

$$F^{(n)}(s) = (-1)^n \int_0^\infty e^{-sx} x^n f(x)\, dx = (-1)^n \mathcal{L}[x^n f].$$

Answers

1. (a) $1/(s - 1)^2$ $(s > 1)$, $2/s^3$ $(s > 0)$, $2/(s - 1)^3$ $(s > 1)$;

(b) $1/(s^2 - 2s + 2)$, $(s - 1)/(s^2 - 2s + 2)$.

2. (a) $\dfrac{a}{s^2 - a^2}$ $(s > |a|)$;

(b) $\dfrac{s}{s^2 - a^2}$ $(s > |a|)$;

(c) $\dfrac{b}{(s - a)^2 + b^2}$ $(s > a)$;

(d) $\dfrac{s - a}{(s - a)^2 + b^2}$ $(s > a)$.

3. (b) $\dfrac{1}{s}\left[\dfrac{1}{1 + e^{-s}}\right]$ $(s > 0)$;

(c) $\dfrac{1}{s^2} - \dfrac{1}{s}\left[\dfrac{1}{e^s - 1}\right]$ $(s > 0)$.

4. (b) $\dfrac{1}{1 + s^2}\coth\left(\dfrac{s\pi}{2}\right)$ $(s > 0)$.

2   Applying the transform

In applying the transform it will be helpful to note that the Laplace transform is what is known as a *linear operator*; that is, if $f(x)$ and $g(x)$ are functions that possess Laplace transforms, then

$$\mathscr{L}[hf(x) + kg(x)] = h\mathscr{L}[f(x)] + k\mathscr{L}[g(x)]   (h, k \text{ constants}).$$

This is an immediate consequence of the fact that

$$\int_0^\infty e^{-sx}[hf(x) + kg(x)]\, dx = h\int_0^\infty e^{-sx}f(x)\, dx + k\int_0^\infty e^{-sx}g(x)\, dx.$$

Thus, for example,

$$\mathscr{L}[3x^2 - 5x + 7] = 3\mathscr{L}[x^2] - 5\mathscr{L}[x] + 7\mathscr{L}[1]$$

$$= 3\left(\frac{2!}{s^2}\right) - 5\left(\frac{1}{s^2}\right) + 7\left(\frac{1}{s}\right)$$

$$= \frac{1}{s^3}(6 - 5s + 7s^2)   (s > 0).$$

Next, we state the following result the proof of which is omitted.

Lemma 2.1. Consider the differential equation

(2.1) $$y^{(n)} + a_1 y^{(n-1)} + \cdots + a_{n-1}y' + a_n y = f(x),$$

where the coefficients a_i are constants, and $f(x)$ is of exponential order. Solutions $y(x)$ and their first n derivatives are of exponential order and hence possess Laplace transforms. Further,

(2.2)
$$\lim_{x \to \infty} y^{(n)}(x)e^{-sx} = 0 \qquad (n = 0, 1, 2, \ldots).$$

if s is any sufficiently large positive constant.

Suppose y is a solution of a differential equation (2.1). Then

$$\mathscr{L}[y'] = \int_0^\infty e^{-sx}y' \, dx,$$

and an integration by parts yields

(2.3)
$$\mathscr{L}[y'] = ye^{-sx}\Big|_0^\infty + s\int_0^\infty e^{-sx}y \, dx$$
$$= s\mathscr{L}[y] - y(0).$$

Similarly,

(2.4)
$$\mathscr{L}[y''] = y'e^{-sx}\Big|_0^\infty + s\int_0^\infty e^{-sx}y' \, dx$$
$$= -y'(0) + s\mathscr{L}[y']$$
$$= s^2\mathscr{L}[y] - sy(0) - y'(0).$$

More generally,

(2.5)
$$\mathscr{L}[y^{(n)}] = y^{(n-1)}e^{-sx}\Big|_0^\infty + s\int_0^\infty e^{-sx}y^{(n-1)} \, dx$$
$$= s\mathscr{L}[y^{(n-1)}] - y^{(n-1)}(0).$$

If we apply formula (2.5) to $\mathscr{L}[y^{(n-1)}]$ and repeat the process, we have, finally, that

(2.6)
$$\mathscr{L}[y^{(n)}] = s^n\mathscr{L}[y] - s^{n-1}y(0) - s^{n-2}y'(0) - \cdots - sy^{(n-2)}(0) - y^{(n-1)}(0).$$

Consider now the differential system

(2.7)
$$y'' + y = x,$$
$$y(0) = 0, \qquad y'(0) = 2.$$

If we denote by $y(x)$ the solution of this system, we have

$$y''(x) + y(x) \equiv x,$$

and the Laplace transforms of each side of this identity must be equal; that is,

$$\mathscr{L}[y''] + \mathscr{L}[y] = \mathscr{L}[x].$$

We have

$$\{s^2 \mathscr{L}[y] - sy(0) - y'(0)\} + \mathscr{L}[y] = \frac{1}{s^2} \qquad (s > 0),$$

or

$$(s^2 + 1)\mathscr{L}[y] = 2 + \frac{1}{s^2}.$$

Thus,

(2.8)
$$\mathscr{L}[y] = \frac{2s^2 + 1}{s^2(s^2 + 1)}.$$

Our problem will be solved if we can determine a function $y(x)$, the Laplace transform of which is given by the right-hand member of (2.8). How shall such a function be found? A table of Laplace transforms is a necessity but, as is the case with integral tables, one cannot expect to find *every* function listed. That in (2.8), in particular, is not likely to be found in such a table. However, writing the right-hand member of (2.8) in terms of partial fractions will be helpful. We have

$$\mathscr{L}[y] = \frac{1}{s^2} + \frac{1}{s^2 + 1}.$$

A reference to the list of transforms in Table (1.9) indicates that

$$\mathscr{L}[y] = \mathscr{L}[x] + \mathscr{L}[\sin x],$$
$$= \mathscr{L}[x + \sin x].$$

We wish to conclude that the solution we seek is $x + \sin x$ (as is actually the case), but perhaps there is more than one function having the same Laplace transform. This question is settled by the following theorem the proof of which lies beyond the scope of this book.

Theorem 2.1. *If $f(x)$ and $g(x)$ are of exponential order and if*
$$\mathscr{L}[f] = \mathscr{L}[g],$$
then
$$f(x) = g(x)$$
at each point where both functions are continuous.

Accordingly,

$$x + \sin x$$

is the solution of (2.7) that we seek.

The procedure we have followed in solving the system (2.7) is typical of the use of Laplace transforms to solve initial value problems—that is, it will be recalled, problems where the conditions on the solution of a differential equation are conditions at a single point, such as

$$y(0) = 0, \qquad y'(0) = 2,$$

in the case of the system (2.7). By algebraic manipulations we determine that the Laplace transform $\mathscr{L}[y(x)]$ of the solution we seek can be written in the form

$$\mathscr{L}[y(x)] = G(s).$$

The problem then becomes: What function $y(x)$ has $G(s)$ as its Laplace transform? This problem is analogous to the problem of formal integration, where to evaluate an integral

$$\int f(x) \, dx$$

we ask what function has $f(x)$ as its derivative. And, like formal integration, the problem of finding a function that has $G(s)$ as its Laplace transform may be very difficult. This process is known as the *inversion* of a Laplace transform. In symbols, one frequently uses the notation $\mathscr{L}^{-1}[G(s)]$. Thus, in the example above one may write

$$\mathscr{L}^{-1}\left[\frac{2s^2 + 1}{s^2(s^2 + 1)} \right] = x + \sin x.$$

Exercises

1.   Use the formulas derived in the text to compute the following Laplace transforms:

(a) $\mathscr{L}[3]$;
(b) $\mathscr{L}[\pi]$;
(c) $\mathscr{L}[2x - 1]$;
(d) $\mathscr{L}[ax^2 + bx + c]$, a, b, and c constants;
(e) $\mathscr{L}[x - \sin x]$;
(f) $\mathscr{L}[3e^x + \cos 2x]$.

2.   Find functions whose Laplace transforms are the following:

(a) $\dfrac{7}{s^2}$;

(b) $\dfrac{60}{s^5}$;

(c) $\dfrac{1}{s^8}$;

(d) $\dfrac{1}{s^3} + \dfrac{1}{s^2 + 1}$;

(e) $\dfrac{1}{s(s^2 + 1)}$;

(f) $\dfrac{2}{s(s - 1)}$.

[*Hint.* Use partial fractions.]   [*Hint.* Use partial fractions.]

Use the method of Laplace transforms to solve the following differential equations subject to the given conditions.

3. $y'' - y = 1 + x$,
     $y(0) = 3$,
     $y'(0) = 0$.

4. $y'' + y = 2$,
     $y(0) = 1$,
     $y'(0) = 0$.

5. $y'' - y' - 2y = \sin x$,
     $y(0) = \frac{1}{10}$,
     $y'(0) = \frac{27}{10}$,

6. $y'' - 3y' + 2y = 2 \cos t$,
     $y(0) = 1$,
     $y'(0) = 1$.

7. $y''' + 4y'' + 3y' = 8e^t$,
     $y(0) = 4$,
     $y'(0) = -3$,
     $y''(0) = 11$.

8. $y''' - y' = 2x + 4e^x$,
     $y(0) = 1$,
     $y'(0) = 6$,
     $y''(0) = 0$.

Answers

1.   (a) $\dfrac{3}{s}$ $(s > 0)$;   (d) $\dfrac{2a}{s^3} + \dfrac{b}{s^2} + \dfrac{c}{s}$ $(s > 0)$;   (e) $\dfrac{1}{s^2} - \dfrac{1}{s^2 + 1}$ $(s > 0)$.

2.   (a) $7x$; (c) $\dfrac{x^7}{7!}$; (e) $1 - \cos x$.

3. $-1 - x + \frac{5}{2}e^x + \frac{3}{2}e^{-x}$

5. $-\frac{3}{10}\sin x + \frac{1}{10}\cos x + e^{2x} - e^{-x}$

7. $e^t + 2 + e^{-3t}$

5

Some applications

1 Newton's laws of motion

Differential equations had their beginnings in mechanics, and some of the most interesting applications of the theory of differential equations are still to be found in this area of applied mathematics. In this chapter we shall examine some simple mechanical and other situations that can be described by differential equations.

We begin with a statement of Newton's laws of motion:

1. *A particle at rest remains at rest, and a particle in motion moves in a straight line with constant velocity unless acted upon by some external force.*

2. *The time rate of change of momentum of a particle is proportional to the resultant of the external forces acting upon the particle.*

3. *Action and reaction are two forces which are equal in magnitude and opposite in sense along the same line of action.*

The second law of motion leads immediately to a differential equation. The momentum of a particle is defined as the product mv, where m is its mass, and v, its velocity. If f denotes the resultant force acting on the particle, Newton's second law states that

(1.1)
$$\frac{d}{dt}(mv) = gf,$$

where g is a constant of proportionality. If the mass is constant, this equation becomes

(1.2)
$$m\frac{dv}{dt} = gf.$$

The quantity $\frac{dv}{dt}$ is, of course, the acceleration of the particle. Thus, if the mass of the particle remains constant, the second law states that the product of mass by acceleration is proportional to the applied force. The value of the constant g depends upon the units employed.

Units. Our purpose is to study differential equations and properties of their solutions. Mathematical analysis is concerned with pure numbers and has little concern for dimensions. Thus, a mathematician looking at the differential equation

$$\frac{dv}{dt} = 6$$

cares little whether the dimension of the number 6 is miles per half-second squared or bacteria per square mile. The solution is the same in either case. Indeed, this illustrates precisely one of the great strengths of mathematics— that, once those aspects of a problem to which mathematical analysis can be applied are abstracted, the solution proceeds independent of the origin of the problem.

Nevertheless, a few comments on units in connection with equation (1.2) may, it is hoped, not add to any confusion that may already exist in the student's mind on this useful but hardly inspiring topic.

In many respects the simplest system of units to employ in connection with equation (1.2) is that in which distance is measured in feet, time in seconds, mass in pounds, and force (weight) in pounds. Then experiment indicates that $g = 32$, approximately, at the surface of the earth. Some persons interested in the applications may wish to insist that $g = 32$ ft./sec.^{2}. In this system, a mass of 1 lb. weighs 1 lb. at the earth's surface.

Suppose we apply these notions to a specific problem before we comment further on this system of units.

Example.   At a point 96 ft. above the ground a ball of mass m is thrown upward with an initial speed of 16 ft./sec. Assuming that the only force acting is that of gravity, find how long it will take the ball to reach the ground and the velocity with which it strikes the ground.

Let us suppose the motion of the ball is along an axis y, distance being measured in a positive sense from the ground to the ball (see Fig. 5.1). Then

FIG. 5.1

velocity in an upward direction will have positive values and velocity toward the ground will have negative values.

Since the only force acting upon the ball is the weight m lb., we have as the differential equation of motion

(1.3)
$$m\frac{dv}{dt} = -gm,$$

or

(1.3)′
$$\frac{dv}{dt} = -32.$$

Note the minus sign in (1.3) and in (1.3)′. This appears because the force is acting downward, and hence tends to decrease v algebraically.

We must solve the *system*

$$\frac{dv}{dt} = -32,$$

(1.4)
$$t = 0, \qquad \frac{dy}{dt} = 16, \qquad y = 96.$$

Conditions such as (1.4) that must be satisfied by a solution of a differential equation are called *initial* or *boundary* or *side* conditions. When the differential equation is combined with its side conditions, the result is frequently called a *differential system*.

From the differential equation in (1.4) we see that

$$v = \frac{dy}{dt} = -32t + c_1 \qquad (c_1 \text{ constant}).$$

When $t = 0$, $v = 16$. It follows that $c_1 = 16$ and that

(1.5) $$v = \frac{dy}{dt} = -32t + 16.$$

From (1.5) we then have

$$y = -16t^2 + 16t + c_2 \qquad (c \text{ constant}).$$

From the condition that $y = 96$, when $t = 0$, we then have $c_2 = 96$ and

$$y = -16t^2 + 16t + 96.$$

The ball will strike the ground when $y = 0$; that is, when

$$-16t^2 + 16t + 96 = 0,$$

or when $t = 3$ or $t = -2$. Since the value $t = -2$ has no meaning in this problem, we conclude that the ball hits the ground in 3 sec. Substituting this value in (1.5) we see that the corresponding velocity is -80 ft./sec. The answer states that the ball is traveling at a speed of 80 ft./sec., when $t = 3$ sec., and the minus sign indicates that its direction of motion at that instant is downward.

If in this example the ball had been thrown *downward* initially at a speed of 16 ft./sec., the differential system to be solved would have been

$$\frac{dv}{dt} = -32,$$

$$t = 0, \qquad \frac{dy}{dt} = -16, \qquad y = 96.$$

The pedagogical objection raised most often to this system of units—the so-called *English* system—is the fact that two quite different quantities, mass and weight, are both measured in *pounds*. And a pound of mass is different conceptually from a pound of force (weight). A pound of mass will not weigh a pound on the surface of the moon.

It is not clear that the advantages of the foregoing system do not offset the slight confusion mentioned above, but engineers frequently prefer to write equation (1.2) in the form

(1.6) $$f = \frac{w}{g}\frac{dv}{dt},$$

where the force f and the weight w are measured in pounds, and g and the acceleration dv/dt are measured in ft./sec.^{2}. Again, $g = 32$ ft./sec.^{2}, approximately, at the earth's surface. An advantage of this approach is that mass does not appear explicitly in this equation. It is, however, determined by the quotient

$$m = \frac{w}{g}.$$

The unit of mass, when w is measured in pounds and $g = 32$, is called a *slug*. Thus, an object that weighs 96 lb. on the surface of the earth has a mass of 3 slugs. On the surface of the moon, for example, both w and g would be about $\frac{1}{6}$ of their values on the earth, but the ratio w/g would be unchanged.

Suppose that a ball that weighs 10 lb. is thrown upward at some point of the earth's surface with an initial velocity of 16 ft./sec. In this case, $w = 10$ lb., $g = 32$ ft./sec.^{2}, $f = 10$ lb., and equation (1.6) becomes

$$-10 = \frac{10}{32}\frac{dv}{dt},$$

or

$$\frac{dv}{dt} = -32.$$

And the motion of the ball may be determined as in the last example. The mass of the ball is, incidentally, $10/32$ slug.

Physicists like to regard Newton's second law as

(1.7) $$f = m\frac{dv}{dt};$$

that is, they choose units so that the constant of proportionality is 1. To accomplish that, they measure f in *dynes*, m in *grams*, and dv/dt in cm./sec.^{2}. On the earth's surface $g = 980$ cm./sec.^{2}, approximately, and $m = w/g$. Or, if mass is measured in pounds, distance in feet, and time in seconds, the force f provided by equation (1.7) is measured in units called *poundals*. Thus, for example, we see from (1.7) that 1 poundal is the force required to give 1 lb. of mass an acceleration of 1 ft./sec.^{2} at the surface of the earth.

Suppose that mass is measured in grams, force in dynes, time in seconds, and distance in centimeters. Equation (1.3)$'$ becomes

(1.8) $$\frac{dv}{dt} = -980,$$

and for the example in the text we would then have

$$t = 0, \quad \frac{dy}{dt} = 487.7\,\frac{\text{cm.}}{\text{sec.}}, \quad y = 2926.1 \text{ cm.}$$

Accordingly,

$$v = \frac{dy}{dt} = -980t + 487.7,$$

$$y = -490t^2 + 487.7t + 2926.1.$$

Then, $y = 0$ when $t = 3.0$ sec., as before. The maximum height the ball will attain occurs when $t = 487.7/980 = 0.50$ sec., at which time $y = 3047.45$ cm. $= 100.0$ ft. And so on.

Exercise

1. If mass is measured in tons, weight in tons, distance in miles, and time in hours, what is the numerical value of g at the earth's surface?

In this text, inasmuch as our objective is to illustrate the application of differential equation theory to certain simple problems in mechanics and not to teach mechanics as such, we shall usually limit our discussion to two systems of units:

1. mass in pounds, force in pounds, distance in feet, time in seconds;
2. mass in slugs, force in pounds, distance in feet, time in seconds.

Example. A particle of mass m slugs is projected with initial velocity v_0 down a frictionless inclined plane whose angle with the horizontal is θ $(0 < \theta < \pi/2)$ (Fig. 5.2). Study the motion of the particle.

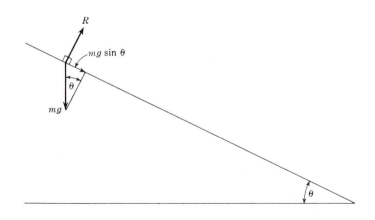

FIG. 5.2

We shall suppose that the only forces acting upon the body are the force of gravity acting vertically (that is, the weight of the particle) and the reaction R of the plane (the line of action of which is perpendicular to the plane). The weight is then numerically equal to mg, so the vertical force is mg lb. The component of this force along the plane is $mg \sin \theta$. Since the component of R parallel to the plane is zero, by Newton's second law we have

$$m \frac{dv}{dt} = gm \sin \theta,$$

or

(1.9)
$$\frac{dv}{dt} = g \sin \theta.$$

Since $g \sin \theta$ is constant, it follows that

$$v = (g \sin \theta)t + c_1,$$

where c_1 is a constant which may be determined from the condition that $v = v_0$ when $t = 0$ (since the initial velocity was given as v_0). Thus, if s is the distance of the particle at time t measured from its initial position down the inclined plane,

(1.10)
$$v = \frac{ds}{dt} = (g \sin \theta)t + v_0,$$

and

$$s = \tfrac{1}{2}(g \sin \theta)t^2 + v_0 t + c_2.$$

Since we have supposed that $s = 0$ when $t = 0$, it follows that the constant of integration c_2 has the value zero.

The motion of the particle is then completely described by the equation

(1.11)
$$s = \tfrac{1}{2}(g \sin \theta)t^2 + v_0 t.$$

Note that equation (1.10) is immediately obtainable from (1.11) by differentiation.

If, for example, $\theta = \dfrac{\pi}{6}$ and $v_0 = 10$ ft./sec., then $g = 32$ ft./sec.², and

$$s = 8t^2 + 10t.$$

After, say, 3 sec. the particle has moved down the plane a distance of 102 ft. from its starting point, and its velocity at this instant is 58 ft./sec.

If the particle were sliding with a coefficient of friction μ, the differential equation of motion would become

(1.12)
$$m \frac{dv}{dt} = mg \sin \theta - \mu mg \cos \theta,$$

since the resultant force acting upon the particle along its path would be $mg \sin \theta$ (the component of its weight along its path) minus $\mu mg \cos \theta$ (by definition, the force of friction acting to retard the motion of the particle along its path). The conditions the solution of (1.12) must satisfy would again be

(1.12)′
$$t = 0, \qquad s = 0,$$
$$t = 0, \qquad v = v_0.$$

As noted above, when the differential equation is combined with side conditions, the result is called a differential system. Thus, equations (1.12) and (1.12)′ lead to the differential system

$$\frac{dv}{dt} = g(\sin \theta - \mu \cos \theta),$$

(1.13)
$$t = 0, \qquad s = 0,$$
$$t = 0, \qquad v = v_0.$$

We readily obtain

(1.14)
$$s = \tfrac{1}{2}g(\sin \theta - \mu \cos \theta)t^2 + v_0 t$$

as the solution of (1.13). Equation (1.14) together with

$$v = \frac{ds}{dt} = g(\sin \theta - \mu \cos \theta)t + v_0$$

enable us to describe the position and the velocity of the particle at time t.

Critique. More precisely, when $\mu \neq 0$, two cases arise in the preceding example: (1) $\mu \leq \tan \theta$, and (2) $\mu > \tan \theta$. In case (1) the solution as given applies without modification. In case (2) it is clear that unless the particle is given an initial velocity there will be no motion.

Example. A particle of mass m leaves the earth with initial velocity v_0 in a direction that makes an angle α with the horizontal. Study the particle's motion, neglecting all forces except that of gravity.

It is convenient to apply Newton's second law to the components of motion as follows. The only force acting is that of gravity, which acts in a vertical direction; consequently, we have (see Fig. 5.3)

$$m\frac{d^2y}{dt^2} = -gm, \qquad m\frac{d^2x}{dt^2} = g \cdot 0,$$

$$t = 0, \quad \frac{dy}{dt} = v_0 \sin \alpha, \qquad t = 0, \quad \frac{dx}{dt} = v_0 \cos \alpha,$$

$$t = 0, \qquad y = 0; \qquad t = 0, \qquad x = 0.$$

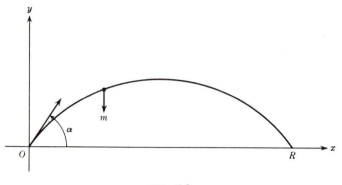

FIG. 5.3

It is easy to see, successively, that

(1.15) $$\frac{dy}{dt} = -gt + v_0 \sin \alpha, \qquad \frac{dx}{dt} = v_0 \cos \alpha,$$

and

(1.16) $$y = -\tfrac{1}{2}gt^2 \quad (v_0 \sin \alpha)t, \qquad x = (v_0 \cos \alpha)t.$$

Thus, equations (1.16) furnish the position of the particle at time t, and equations (1.15) provide components of its velocity at that instant.

Exercises

In the first six exercises set up the differential equations of motion of the given particles. Solve these equations, determining the constants of integration by use of suitable boundary conditions.

1. A missile is projected downward from a balloon 960 ft. above the earth. The initial speed of the missile is 256 ft./sec. When and with what velocity will the missile strike the earth? (Neglect all forces except that of gravity.)

2. Answer the same question and assume the same conditions given in Exercise 1, except that the missile is projected upward with an initial speed of 256 ft./sec.

3. A ball leaves the earth at an angle of $\pi/6$ and an initial speed of 30 ft./sec. When will it strike the ground? What will be its speed in its path at the instant of impact? How far will it then be from its starting point?

4. The same as Exercise 3 when the angle of launch is $\pi/3$ and the initial speed is 90 ft./sec.

5. A frictionless plane is inclined at an angle of $\pi/6$. A block is projected down the plane with an initial speed of 20 ft./sec. How far will the block have traveled after 4 sec.? At what velocity will it be traveling?

6. The same as Exercise 5, when the block is projected upward with an initial speed of 30 ft./sec.

7. Do Exercise 5 when the plane has a coefficient of friction of 0.15 with the block.

8. A block is projected up a plane that makes an angle of $\pi/4$ with the horizontal. The coefficient of friction between the plane and the block is 0.15, and the initial speed of the block is 30 ft./sec. Where will the block be at the end of 1 sec.? What will be its velocity? Where will it be after 3 sec.? What will its velocity be then? (Assume that the coefficient of "starting" friction is also 0.15.)

9. A particle initially of mass m_0 is projected upward from the surface of the earth with an initial speed of 32 ft./sec. It loses mass at the rate of $\frac{1}{2}m$ lb./sec., where m is the mass at time t. When and with what velocity will the particle strike the earth? (Assume that the only force acting is that of gravity and that Newton's second law of motion applies as stated.)

   (*Note.* It will be helpful to employ some sort of approximate method in solving for the time of impact.)

10. Show that the path of the particle in the last example in the text is a parabola. Find the range OR. What value of α makes the range a maximum?

11. A sled has an initial velocity of 30 mi./hour on a level ice field. It comes to rest in 20 min. What is the coefficient of friction between the sled and the ice?

12.* Study the motion of a particle on a plane which makes an angle of $\pi/6$ with the horizontal when $\mu = 0.6$ and the initial speed is 10 ft./sec. up the plane.

13. Assume that on the moon's surface $g = 16.3$ ft./sec.^{2}. At a point 96 ft. above that surface a ball of mass 12 slugs is thrown upward with an initial speed of 16 ft./sec. Assuming that the only force acting is the force of gravity on the moon, find how long it will take the ball to reach the ground. How high will it rise?

14. Solve the same problem as in Exercise 13 when the motion takes place on a planet on the surface of which $g = 64$ ft./sec.^{2}.

Answers

1. 3.14 sec.; -356.34 ft./sec.

2. 19.14 sec.; -356.34 ft./sec.

3. 0.94 sec.; 30 ft./sec.; 24.42 ft.

4.   4.87 sec.; 90 ft./sec.; 219.21 ft.

5.   208 ft.; 84 ft./sec.

6.   64.2 ft.; 34.0 ft./sec.

7.   174.7 ft.; 67.4 ft./sec.

8.   17.0 ft.; 4 ft./sec.; 1.8 ft. above original starting point; -35.5 ft./sec.

9.   2.51 sec.; -48.25 ft./sec.

10. $(v_0^2 \sin 2\alpha)/g$; $\pi/4$.

11.   0.0011.

13.   4.55 sec.; 103.85 ft.

14.   2 sec.; 98 ft.

2   Gravitational attraction

Newton's celebrated law of gravitation asserts that *every particle of matter in the universe attracts every other particle, with a force whose direction is that of the line joining the two, and whose magnitude is directly as the product of their masses, and inversely as the square of their distance from each other.*[†]

Example.   Study the motion of a particle falling to the earth from a great distance. Neglect all forces except the gravitational force between the two bodies.

Let the mass of the particle be m and that of the earth (assumed to be a homogeneous sphere) be M. Then the force acting upon the particle is given by

$$f = \frac{kmM}{r^2},$$

where r is the distance of the particle from the center of the earth[‡] and k is a constant of proportionality. When $r = R$, the radius of the earth, f is the weight m of the particle at the earth's surface. Thus, $k = \dfrac{R^2}{M}$, and

$$f = \frac{mR^2}{r^2} \text{ lb.}$$

[†] This statement of the law is taken from O. D. Kellogg, *Foundations of Potential Theory*, Springer-Verlag (1929). Kellogg credits Thomson and Tait, *A Treatise on Natural Philosophy*, Cambridge University Press, New York (1912), with this formulation.

[‡] It may be shown that it is a correct extension of the law of gravitation in this situation to treat the mass of the earth as though it were concentrated at the center (see Kellogg, *op. cit.*).

Newton's second law of motion may now be invoked, and we have as the differential equation of motion of the particle

$$m\frac{d^2r}{dt^2} = -gm\frac{R^2}{r^2},$$

or

(2.1)
$$\frac{d^2r}{dt^2} = -g\frac{R^2}{r^2}.$$

We multiply both members of equation (2.1) by $2\frac{dr}{dt}$, and integration yields at once

(2.2)
$$\left(\frac{dr}{dt}\right)^2 = \frac{2gR^2}{r} + c_1.$$

Let us set $r = r_0$ and $\frac{dr}{dt} = 0$, when $t = 0$. Then

$$c_1 = -\frac{2gR^2}{r_0},$$

and (2.2) becomes

(2.3)
$$\left(\frac{dr}{dt}\right)^2 = 2gR^2\left(\frac{1}{r} - \frac{1}{r_0}\right).$$

Thus, the velocity v_R of the particle at the earth's surface is given by

(2.4)
$$v_R^2 = 2gR^2\left(\frac{1}{R} - \frac{1}{r_0}\right).$$

If r_0 is very large, v_R is approximately $\sqrt{2gR}$, which is about 7 mi./sec.†
    Equation (2.3) leads at once to

(2.5)
$$\frac{dr}{dt} = -\sqrt{\frac{2gR^2}{r_0}}\frac{\sqrt{r_0r - r^2}}{r}.$$

The minus sign is employed since r decreases as t increases. The variables are separable in equation (2.5), and we have at once

(2.6)
$$\frac{r\,dr}{\sqrt{r_0r - r^2}} + \sqrt{\frac{2gR^2}{r_0}}\,dt = 0 \qquad (0 < r < r_0).$$

An integration yields

$$t = \sqrt{\frac{r_0}{2gR^2}}\left(\sqrt{r_0r - r^2} - \frac{r_0}{2}\sin^{-1}\frac{2r - r_0}{r_0}\right) + c_1.$$

---

† This is, then, the velocity of escape from the earth's gravitational field, if the resistance of air is neglected.

Since, as $r \to r_0$, $t \to 0$, we must have $c_1 = \sqrt{\dfrac{r_0}{2gR^2}} \dfrac{\pi r_0}{4}$, and finally

$$t = \sqrt{\frac{r_0}{2gR^2}} \left(\sqrt{r_0 r - r^2} - \frac{r_0}{2} \sin^{-1} \frac{2r - r_0}{r_0} + \frac{\pi r_0}{4} \right).$$

The time required for the particle to fall to the earth's surface is

$$t = \sqrt{\frac{r_0}{2gR^2}} \left(\sqrt{r_0 R - R^2} - \frac{r_0}{2} \sin^{-1} \frac{2R - r_0}{r_0} + \frac{\pi r_0}{4} \right).$$

The student will recall that if t is measured in seconds and r and r_0 in feet, then $g = 32$.

3   Simple harmonic motion

If a particle moves along the x-axis according to the law

(3.1)
$$m \frac{d^2x}{dt^2} = -kx,$$

where k is a positive constant, it is said to be in *simple harmonic motion* in its path. Equation (3.1) may be written in the form

(3.2)
$$\frac{d^2x}{dt^2} + a^2 x = 0 \qquad (a^2 > 0).$$

The general solution of (3.2) is seen to be

$$x = c_1 \sin at + c_2 \cos at,$$

the *period* of which is

$$T = \frac{2\pi}{a},$$

and the *amplitude* of which is $\sqrt{c_1^2 + c_2^2}$.

Exercises

1. A particle moves along the x-axis in simple harmonic motion, and it is known that at $t = 0$, $x = 0$ and $\dfrac{dx}{dt} = 1$. Given that the particle's period is 2π, discuss its motion.

2. A point moves on the circumference of a circle at constant angular speed. Show that the vertical projection of the point on the x-axis is in simple harmonic motion.

3. A point moves on the x-axis according to the law

$$\frac{d^2x}{dt^2} + 6\frac{dx}{dt} + 25x = 0,$$

$$x(0) = 0, \qquad \dot{x}(0) = 2.$$

Discuss the motion as t varies from 0 to $+\infty$.

4. A point moves along the x-axis according to the law

$$\frac{d^2x}{dt^2} + \frac{dx}{dt} = 0,$$

$$x(0) = 2, \qquad \dot{x}(0) = -1.$$

Discuss the motion as t varies from 0 to $+\infty$.

5. A point moves along the positive y-axis according to the law

$$\frac{d^2y}{dt^2} + 3\frac{dy}{dt} + 2y = 0,$$

$$y(0) = 0, \qquad \dot{y}(0) = 10.$$

Discuss the motion $(0 \leq t < \infty)$.

6. It can be shown that if a hole is bored through the center of the earth, the attraction on a particle in the resulting tunnel is proportional to the distance of the particle from the center of the earth. (Air resistance is neglected.) A stone is dropped into the tunnel. Discuss its motion.

7. Show that the differential equation of motion of a simple pendulum bob is $\frac{md^2s}{dt^2} = -mg \sin \theta$. Note that $s = l\theta$, and thus obtain the differential equation

$$\frac{d^2\theta}{dt^2} + \frac{g}{l} \sin \theta = 0.$$

Discuss the motion when $\sin \theta$ above is replaced by θ (a good approximation when $|\theta|$ is small). See Fig 5.4.

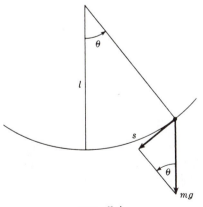

FIG. 5.4

4   The vibrating spring

In this section we study the motion of a body B which is fastened to the lower end of a vertical coiled spring, the upper end of which is rigidly secured (see Fig. 5.5). Let B be lowered carefully to the point O of equilibrium,

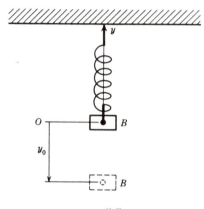

FIG. 5.5

that is, the point at which it will remain at rest. Suppose B is given a small vertical displacement of y_0 ft. and released with an initial vertical velocity v_0. We shall suppose that the force exerted by the spring on B has the following properties:

1. It acts along a vertical line through the center of gravity of B and is directed from the position of B toward O, its point of equilibrium.
2. Its magnitude at any instant t is proportional to the difference $|l - l_0|$, where l is the length of the spring at time t, and l_0 is its length when B is at the point of equilibrium.

The constant of proportionality in property 2 is known as the *spring constant*.

It is clear that if we neglect any other forces acting on B (such as air resistance), we may analyze the motion of B as follows. Since the motion of B occurs along a vertical line, we suppose that it takes place along a y-axis oriented as usual with its origin at O. Velocities and accelerations directed upward will be regarded as being of positive sign and those directed downward as being negative. Suppose we measure force in pounds, mass in pounds,

distance in feet, and time in seconds. We then have as the differential
system expressing the motion of B:

$$m\frac{d^2y}{dt^2} = -g(ky),$$

(4.1)

$$t = 0, \qquad y = y_0, \qquad \frac{dy}{dt} = v_0,$$

where k is the spring constant and g is approximately 32.

Example. A spring is stretched 4 in. by a 6-lb. weight. Let a 16-lb.
weight B be attached to the spring and released 3 in. below the point of
equilibrium with an initial velocity of 2 ft./sec. directed downward. Describe
the motion of B.

The first sentence in the statement of the problem enables us to determine
the spring constant k. We have

$$f = k(l - l_0),$$

or

$$6 = k(\tfrac{1}{3}).$$

Thus, $k = 18$, and equations (4.1) become

$$\frac{d^2y}{dt^2} + 36y = 0,$$

(4.1)'

$$t = 0, \qquad y = -\tfrac{1}{4}, \qquad \frac{dy}{dt} = -2.$$

It is readily seen that the solution of this system is

$$y = -\tfrac{1}{3}\sin 6t - \tfrac{1}{4}\cos 6t.$$

The amplitude of this motion is

$$\sqrt{\tfrac{1}{9} + \tfrac{1}{16}} = \tfrac{5}{12} \text{ ft.},$$

and its period is

$$T = \frac{2\pi}{6} = \frac{\pi}{3} \text{ sec.}$$

Suppose, however, that we are using the International System of Units
which measures force in newtons, mass in kilograms, distance in meters, and
time in seconds. The differential system becomes

$$m\frac{d^2y}{dt^2} = -g(ky),$$

(4.1)'

$$t = 0, \qquad y = y_0, \qquad \frac{dy}{dt} = v_0,$$

just as before. Now suppose that a spring is stretched 10 cm. by a mass weighing 25 newtons (25 N). Suppose a 50-N weight is attached to the spring and released 7 cm. below the point of equilibrium with an initial velocity of 63 cm./sec. directed downward.

In this example, $g = 9.8$ m./sec.$^2 = 980$ cm./sec.^{2} (approximately), and we have first that

$$25 = k(10),$$

or $k = 2.5$. The differential equation then becomes

$$\frac{d^2y}{dt^2} + 49y = 0.$$

The boundary conditions in (4.1)' lead to the solution

$$y(t) = -9 \sin 7t - 7 \cos 7t.$$

The amplitude is now $\sqrt{130}$ cm., and the period is $T = 2\pi/7$ sec.

Exercises

1. A spring is stretched 3 in. by an 8-lb. weight. Suppose a 4-lb. weight B is attached to the spring and released 6 in. below the point of equilibrium with an initial velocity of 3 ft./sec. directed downward. Study the motion of B.

2. Assume the same conditions given in Exercise 1, but with the point of release 4 in. above the point of equilibrium and an initial velocity of 2 ft./sec. directed downward. Study the motion of B.

3. Assume the same conditions given in Exercise 1, but with the point of release 6 in. above the point of equilibrium and an initial velocity of 0 ft./sec. Study the motion of B.

4. A spring is such that it would be stretched a ft. by a weight of b lb. Suppose a weight of c lb. is attached to the spring and released at a point h ft. above the point of equilibrium with an initial velocity of v_0 ft./sec. Describe the motion.

5. A spring is stretched 10 cm. by a weight of 18 N. Suppose a 9-N weight B is attached to the spring and released 12 cm. below the point of equilibrium with an initial velocity of 98 cm./sec. directed downward. Study the motion of B.

6. Assume the same conditions as in Exercise 5, but with the point of release 10 cm. above the point of equilibrium and an initial velocity of 28 cm./sec. directed downward. Study the motion of B.

7. Assume the same conditions as in Exercise 5, but with the point of release 5 cm. above the point of equilibrium and an initial velocity of 0 cm./sec.

Answers

1. $-\frac{1}{2}\cos 16t - \frac{3}{16}\sin 16t.$

2. $\frac{1}{3}\cos 16t - \frac{1}{8}\sin 16t.$

3. $\frac{1}{2}\cos 16t.$

4. $h\cos pt + (v_0/p)\sin pt,\ p = \sqrt{32b/ac}.$

5. $-\frac{1}{7}\sin 14t - 12\cos 14t.$

6. $-2\sin 14t + 10\cos 14t.$

7. $5\cos 14t.$

5   Damped vibrations

If account is taken of the fact that the vibration of the weight B on the spring discussed in the previous section usually occurs in a fluid, such as air or water or oil or molasses, it is realistic to recognize that the resistance of the fluid retards the motion of B. In many instances it has been found that the retarding or *damping* forces may be approximated by a quantity proportional to the velocity and oppositely directed. The differential equation of motion then becomes

$$(5.1) \qquad m\frac{d^2y}{dt^2} = g\left(-ky - c\frac{dy}{dt}\right),$$

where the positive constant c is an empirical constant determined by experiment.

Equation (5.1) may be written in the form

$$(5.1)' \qquad \frac{d^2y}{dt^2} + 2\alpha\frac{dy}{dt} + \beta^2 y = 0,$$

where

$$2\alpha = \frac{cg}{m} > 0, \qquad \beta^2 = \frac{gk}{m}.$$

The indicial equation associated with equation (5.1)′ is

$$m^2 + 2\alpha m + \beta^2 = 0,$$

the roots of which are

$$m = -\alpha \pm \sqrt{\alpha^2 - \beta^2}.$$

We note that the solutions of (5.1)′ are oscillatory if and only if

$$\beta^2 > \alpha^2.$$

When $\beta^2 = \alpha^2$, the system is said to be *critically* damped.

Exercise

1. Find the value of c that would cause the motion of the spring of the example in Section 4 to be critically damped. When $c = 6$, find the solution $y(t)$ of the differential equation for which $y(0) = -\frac{1}{4}$ and $\dot{y}(0) = -2$, and graph $y(t)$ as a function of t.

Answer

1. $c = 6$.

6 Forced vibrations

Frequently, in addition to the forces already considered in our study of the vibration of a spring, we may also have to deal with an *impressed* force acting vertically upon the weight B, such as might be induced by the vibration of the "rigid" support to which the spring is attached. If we suppose that such a force is a function of the time t, the differential equation of motion may be given the form

(6.1) $$\frac{d^2y}{dt^2} + 2\alpha\frac{dy}{dt} + \beta^2 y = f(t).$$

The situation in which $f(t) = c_1 \sin \omega t + c_2 \cos \omega t$, where c_1, c_2, and ω are given constants, is of especial importance in practical applications.

Exercises

Solve the following differential systems. Describe the behavior of $y(t)$, as $t \to +\infty$.

1. $\dfrac{d^2y}{dt^2} + 2\dfrac{dy}{dt} + y = \sin t, \qquad t = 0, y = 1, \dfrac{dy}{dt} = 2.$

2. $\dfrac{d^2y}{dt^2} + 2\dfrac{dy}{dt} + 2y = \cos t, \qquad t = 0, y = -1, \dfrac{dy}{dt} = -2.$

3. $\dfrac{d^2y}{dt^2} + 4y = -2 \sin 2t, \qquad t = 0, y = 0, \dfrac{dy}{dt} = 1.$ (The behavior of the amplitudes in the vibration characterized by this differential system is termed *resonance*.) Graph $y(t)$.

4. $\dfrac{d^2y}{dt^2} + 3\dfrac{dy}{dt} + 2y = \cos t, \qquad t = 0, y = 0, \dfrac{dy}{dt} = 1.$

5. $\dfrac{d^2y}{dt^2} + 6\dfrac{dy}{dt} + 25y = \sin t, \qquad t = 0, y = 1, \dfrac{dy}{dt} = 0.$

6. $\dfrac{d^2y}{dt^2} + y = -\cos t, \qquad t = 0, y = 0, \dfrac{dy}{dt} = -1.$ \qquad Graph $y(t)$.

Answers

1. $\frac{3}{2}e^{-t} + \frac{7}{2}te^{-t} - \frac{1}{2}\cos t.$

3. $\frac{1}{4}(1 + 2t)\cos 2t.$

5. $e^{-3t}(\frac{103}{102}\cos 4t + \frac{305}{408}\sin 4t) + \frac{2}{51}\sin t - \frac{1}{102}\cos t.$

7   The deflection of beams

Imagine a carpenter carrying a long piece of lumber on his shoulder. That piece of lumber is a *beam*, and it is *deflected*—that is, it bends downward in front of and behind the carpenter's shoulder. Imagine two men picking up a length of "two-by-four," one man at each end—the two-by-four will sag in the middle. That too is a deflected beam. If the two-by-four were *very* long, an effort to pick it up by the ends would result in its breaking—the *elastic limit* of the beam would have been exceeded.

In this section we shall be concerned with the deflection of beams occurring well within their elastic limits. Roughly speaking, we shall seek equations of the curves into which beams are bent by the force of gravity when they are supported at one or both ends.

First let us describe the kind of beam we shall be considering. The beam will have an axis of symmetry AB running the length of the beam (see Fig. 5.6). Every plane perpendicular to the axis of symmetry will have the same

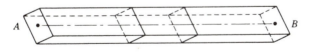

FIG. 5.6

cross-section. We shall suppose that there is a vertical plane through the axis of symmetry with respect to which the beam is symmetric. Finally, we assume that the beam is made of homogeneous material.

When such a beam is picked up by the ends, the axis of symmetry will be bent as shown by the dotted line in Fig. 5.7. It is the equation of this so-called

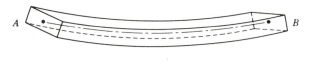

FIG. 5.7

elastic curve which we seek. To that end, we take an x-axis through the original axis of symmetry, putting the endpoint A at the origin. The y-axis is taken as indicated in Fig. 5.8, where we have also indicated the elastic curve when the beam is supported at both ends.

Call the length of the beam b, and choose a point x $(0 < x < b)$ on OB. Next, mark the corresponding point $P(x, y)$ on the elastic curve. If k is the weight of a unit length of the beam (which we have supposed uniform), half the weight is supported at each end of the beam; that is, there is a force of $\dfrac{kb}{2}$ lb. directed upward at O and at B.

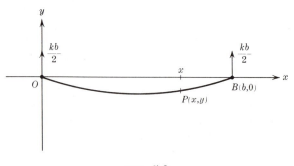

FIG. 5.8

Consider next the first moments about x of the forces to the left of x. The moment due to the force $\dfrac{kb}{2}$ at O is

$$\frac{kb}{2}\, x,$$

while the weight of the segment Ox, which is kx, exerts a moment in the opposite direction in the amount of

$$kx\,\frac{x}{2}.$$

The algebraic sum of these moments is

(7.1) $$M(x) = \frac{kbx}{2} - \frac{kx^2}{2},$$

where we have regarded downward forces as producing negative moments and upward forces as producing positive moments. The student may verify that we would have obtained the same result if we had considered the segment xB to the right of x.

Next, we borrow a result from the theory of the strength of materials which states that

$$(7.2) \qquad\qquad M(x) = EIy'',$$

where E is *Young's modulus*, a constant depending on the material in the beam, and I is an average moment of inertia of a cross-section of the beam with respect to the line parallel to the x-axis through the center of gravity of the cross-section. The product EI is known as the *flexural rigidity* of the beam and is assumed for our present purposes to be a constant. We combine equation (7.2) with (7.1), obtaining the differential equation

$$(7.3) \qquad\qquad EIy'' = \frac{kbx}{2} - \frac{kx^2}{2}$$

for the elastic curve.

We wish to find a solution $y(x)$ of this differential equation subject to the conditions

$$y(0) = y(b) = 0.$$

This is easily computed to be

$$(7.4) \qquad\qquad y(x) = \frac{-k}{24EI} (x^4 - 2bx^3 + b^3x).$$

The maximum deflection of the beam occurs at $x = \dfrac{b}{2}$ (from the symmetry —or the calculus). It is found to be

$$\frac{-5kb^4}{384EI}.$$

The beam of Fig. 5.8, which is supported at both ends, is known as a *simply supported beam*.

Cantilever beams.   When a beam is rigidly supported at one end and the other end is free to move (Fig. 5.9), the beam is called a *cantilever beam*. The axis of symmetry is shown along an x-axis in Fig. 5.10, and the corresponding elastic curve is dotted on the graph.

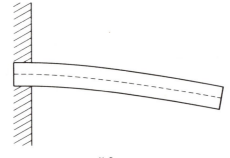

FIG. 5.9

Let x be a point of OB ($0 < x < b$), and compute the first moment of xB about the point x. Its weight is $k(b - x)$, and its first moment (downward) is

$$-k(b - x)\frac{b - x}{2}.$$

Again we set

$$EIy'' = -\frac{k}{2}(b - x)^2,$$

and we wish to find a solution of this differential equation subject to the conditions

$$y(0) = y'(0) = 0.$$

This solution is readily computed to be

$$y(x) = \frac{k}{24EI}(x^4 - 4bx^3 + 6b^2x^2).$$

The deflection of the free end is found by setting $x = b$. It is

$$\frac{kb^4}{8EI}.$$

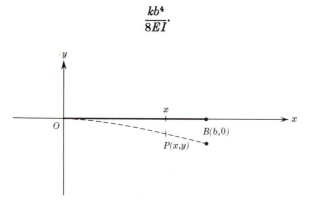

FIG. 5.10

Exercises

1. Suppose a cantilever beam has negligible weight and that a weight of wN is hung from the free end. Find an equation of the elastic curve and the deflection of the free end.

2. Suppose the cantilever beam of Exercise 1 weighs kN per unit of length. Find an equation of the elastic curve and the deflection of the free end.

3. Suppose the simply supported beam of the text has a weight of wN suspended at the midpoint. Find an equation of the elastic curve and its maximum deflection.

Answers

1. $y = \dfrac{wx^2}{6EI}(3b - x); \dfrac{wb^3}{3EI}.$

2. $y = \dfrac{wx^2}{6EI}(3b - x) + \dfrac{kx^2}{24EI}(6b^2 - 4bx + x^2); \dfrac{wb^3}{3EI} + \dfrac{kb^4}{8EI}.$

3. $y = \dfrac{-k}{24EI}(x^4 - 2bx^3 + b^3x) + \dfrac{wx}{12EI}(x - b)(x - 2b).$

8 Suspended cables

Consider a flexible, inextensible cable or rope suspended from two points A and B, not necessarily at the same level, which hangs at rest under the force of gravity (see Fig. 5.11). Take x- and y-axes as shown in Fig. 5.11, the y-axis being perpendicular to the curve of AB at its minimum point V. We wish to find a differential equation of the curve AB.

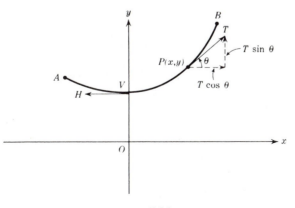

FIG. 5.11

Let $P(x, y)$ be an arbitrary point of the curve, and consider the arc VP. This segment will be in equilibrium under the force of tension T at P, the tension H at V, and the vertical loading W on VP. The tension T and the horizontal force H act along tangents to the curve at P and V, respectively. We regard the vertical loading W on VP as being due to the weight of the cable plus any other weights supported by the cable. Thus, W will vary from point to point.

We resolve T into a horizontal component $T \cos \theta$ and a vertical component $T \sin \theta$. If the density of the load is denoted by $\delta(x)$, then the total vertical load W on the segment VP of the curve is given by the integral

$$(8.1) \qquad\qquad W = \int_0^x \delta(x)\, dx.$$

Since the arc VP is in equilibrium, the algebraic sum of the horizontal and of the vertical forces acting must in each case be zero. Accordingly,

$$T \cos \theta - H = 0,$$
$$T \sin \theta - W = 0.$$

Elimination of T between these equations yields

$$\tan \theta = \frac{W}{H}.$$

But since $\tan \theta = \dfrac{dy}{dx}$ we have

$$(8.2) \qquad\qquad \frac{dy}{dx} = \frac{W}{H}$$

as a differential equation of the curve of AB. We note that H is a constant. As noted above, W is a function of x. It will be useful to differentiate both members of (8.2). We have

$$(8.3) \qquad\qquad \frac{d^2y}{dx^2} = \frac{1}{H}\frac{dW}{dx}.$$

Example. Suppose the weight of a suspended cable is negligible and that it supports a uniform roadway as indicated in Fig. 5.12 (this is the situation in determining the shape of a cable for a suspension bridge, when the weights of the cable and vertical struts are very small compared with the weight of the roadway). By a uniform roadway is meant one in which the weight per linear foot is constant; that is, $\delta(x)$ is a constant δ, and from (8.1) we have

$$\frac{dW}{dx} = \delta.$$

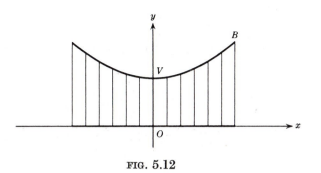

FIG. 5.12

Equation (8.3) becomes

(8.4)
$$y'' = \frac{\delta}{H}.$$

If the coordinates of V are $(0, c)$, we desire a solution of (8.4) subject to the boundary conditions

$$y(0) = c, \qquad y'(0) = 0.$$

The desired solution is readily seen to be the parabola

(8.5)
$$y = \frac{\delta}{2H} x^2 + c.$$

Suppose, in particular, that the dimensions of the suspension bridge are as indicated in Fig. 5.13. In this case, when $x = 0$, $y = 30$, and when $x = 100$, $y = 50$; consequently, the coefficients in (8.5) can be determined, and the parabola has the equation

$$y = 0.002x^2 + 30.$$

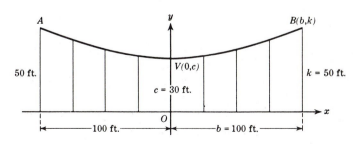

FIG. 5.13

Example.   Suppose the cross-section of the cable in Fig. 5.11 is uniform, and that the only load on the cable is due to its own weight. In this case, the change in weight per unit of *arc length* is constant; that is,

$$\frac{dW}{ds} = k,$$

where k is the weight of a unit length of the cable. But

$$\frac{dW}{ds} = \frac{dW}{dx}\frac{dx}{ds};$$

hence,

$$\frac{dW}{dx} = k\frac{ds}{dx} = k\sqrt{1 + \left(\frac{dy}{dx}\right)^2}$$

Equation (8.3) becomes

(8.6) $$y'' = \frac{k}{H}\sqrt{1 + y'^2}.$$

We note that neither of the quantities x and y appears in this equation. If we set $p = y'$, we have

$$p' = \frac{k}{H}\sqrt{1 + p^2},$$

or

$$\frac{dp}{\sqrt{1 + p^2}} = m\,dx \qquad \left(m = \frac{k}{H}\right).$$

The variables are separated in this differential equation. An integration yields

$$\ln\left(p + \sqrt{1 + p^2}\right) = mx + C_1.$$

We shall take $y = c$ and $p = y' = 0$, when $x = 0$; thus, $C_1 = 0$, and we have

$$p + \sqrt{1 + p^2} = e^{mx}.$$

It follows that

$$1 + p^2 = (e^{mx} - p)^2 = e^{2mx} - 2pe^{mx} + p^2.$$

Accordingly,

$$p = \frac{dy}{dx} = \frac{1}{2}(e^{mx} - e^{-mx}),$$

and, recalling that $y = c$ when $x = 0$, we obtain

(8.7) $$y = \frac{1}{2m}(e^{mx} + e^{-mx}) + \left(c - \frac{1}{m}\right).$$

It is customary to choose the x-axis so that

$$c = \frac{1}{m}.$$

Equation (8.7) becomes

(8.8) $$y = \frac{1}{2m} (e^{mx} + e^{-mx}) \qquad \left(m = \frac{k}{H}\right),$$

which can be written

$$y = \frac{1}{m} \cosh mx.$$

The curve represented by equation (8.8) is called a *catenary*. It is of considerable interest to scientists and engineers as well as to mathematicians. It may be shown, for example, by means of the calculus of variations, that if an arc of a curve AB, as in Fig. 5.14, is rotated about the x-axis, the area of the surface of revolution so generated will be smallest when AB is a segment of a catenary (provided A and B are not too far apart!).

Suppose the points A and B of the catenary (8.8) are on the same level and that A and B are $2b$ units apart. If the slope of the catenary at B is $\lambda\ (>0)$, then

(8.9) $$m = \frac{1}{b} \ln (\lambda + \sqrt{\lambda^2 + 1}).$$

The demonstration of (8.9) is left as an exercise for the student.

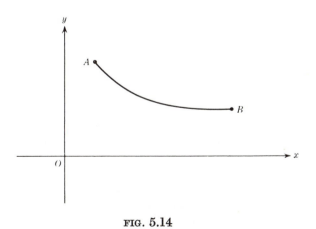

FIG. 5.14

Exercises

1. Derive equation (8.9).

2. Refer to equation (8.9) and find an equation of the catenary when the slope at B is $\frac{3}{4}$, and A and B are 100 units apart.

3. Refer to Fig. 5.13, and suppose that $c = 40$ ft., $k = 60$ ft., $b = 100$ ft. Determine an equation of the cable. Plot the curve on the interval $-100 \le x \le 100$.

4. Solve as in Exercise 3 when $c = 50$ ft., $k = 80$ ft., $b = 100$ ft.

5. A cable weighs $\frac{1}{2}$ lb. per foot. It is suspended from two supports which are at the same level and are 200 ft. apart. The slope of the cable at one support is $\frac{5}{12}$. Find an equation of the curve of the cable. What is the tension at the vertex?

6. The cable of a suspension bridge is of negligible weight compared with the weight of the roadway it supports. The cable supports are at the same level $2b$ ft. apart. The supports are k ft. above the roadway, and the lowest point of the cable is c ft. above the roadway. Suppose the roadway of the bridge weighs δ lb. per foot. Find the tension at the vertex and at a support.

Answers

2. $y = \dfrac{25}{\ln 2} (2^{x/50} + 2^{-x/50})$.

3. $y = 0.002x^2 + 40$.

5. $y = \dfrac{50}{\ln (3/2)} \left[\left(\dfrac{3}{2}\right)^{x/100} + \left(\dfrac{3}{2}\right)^{-x/100} \right]; \dfrac{50}{\ln (3/2)}$.

6. $H = \dfrac{\delta b^2}{2(k - c)}; \; T = \dfrac{\delta b}{2(k - c)} \sqrt{b^2 + 16(k - c)^2}$.

9   Simple electric circuits

In this section we study the use of second-order linear differential equations in describing the flow of electricity in a simple circuit. Such a circuit is shown schematically in Fig. 5.15. In this circuit an electromotive force E causes electricity to flow in a circuit of *inductance* L against a resistance R, depositing an electrostatic charge on one plate of a *capacitor* C. (A capacitor consists of two plates separated by an insulating material. When an electro-

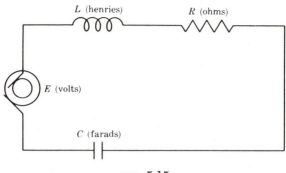

FIG. 5.15

static charge is deposited on one plate of a capacitor, an equal charge opposite in sign appears on the other plate.) Quantity of electricity q is measured in *coulombs*. The *current i* is the rate of flow of electricity; that is,

$$(9.1) \qquad i = \frac{dq}{dt},$$

where i is measured in *amperes*, q in coulombs, and t in seconds.

Electrical inductance is analogous to mass in a mechanical system, and resistance is the analogue of friction. The standard unit of inductance is the *henry* and that of resistance is the *ohm*. The capacity of a capacitor is measured in *farads*. When an inductance L, resistance R, and a capacitance C (L, R, C constants) are connected in series, as shown in Fig. 5.15, it is a consequence of what are known as Kirchhoff's laws† that

$$(9.2) \qquad L\frac{d^2q}{dt^2} + R\frac{dq}{dt} + \frac{1}{C}q = E.$$

Differentiation of (9.2) with respect to t yields

$$(9.3) \qquad L\frac{d^2i}{dt^2} + R\frac{di}{dt} + \frac{1}{C}i = \frac{dE}{dt},$$

inasmuch as $i = \dfrac{dq}{dt}$.

Example. Suppose that a capacitor of capacity 0.01 farad is being charged by a constant electromotive force of 100 volts through an inductance of 0.02 henry against a resistance of 10 ohms. In this case, equation (9.2) becomes

$$(9.2)' \qquad 0.02\frac{d^2q}{dt^2} + 10\frac{dq}{dt} + 100q = 100.$$

---

† An exposition of these laws may be found in most elementary physics textbooks.

Let us suppose further than when $t = 0$, then $q = 0$ and $i = 0$. We have to solve the second-order linear differential equation $(9.2)'$ with constant coefficients subject to the boundary conditions $t = 0$, $q = 0$, $i = 0$.

We note that a particular solution of $(9.2)'$ is $q = 1$, and that the corresponding homogeneous equation has indicial roots which are the solution of the quadratic equation

(9.4) $$0.02m^2 + 10m + 100 = 0.$$

Roots of equation (9.4) are seen to be

$$m = \frac{-5 \pm \sqrt{23}}{0.02} = 50(-5 \pm \sqrt{23}),$$

or

$$m = -10, \ -490 \ \text{(approximately)}.$$

The general solution of equation $(9.2)'$ is, then,

$$q = 1 + c_1 e^{-10t} + c_2 e^{-490t}.$$

Applying the boundary conditions, we have $c_1 = -\frac{49}{48}$, $c_2 = \frac{1}{48}$; hence,

(9.5) $$q = 1 - \tfrac{49}{48}e^{-10t} + \tfrac{1}{48}e^{-490t},$$

and

(9.6) $$i = \frac{dq}{dt} = \frac{490}{48}(e^{-10t} - e^{-490t}).$$

Graphs of q and i as functions of t are given in Figs. 5.16 and 5.17. It will be observed that the charge on the capacitor rapidly approaches 1 coulomb and that the current rises quickly to a peak and then decreases rapidly (see Exercise 1).

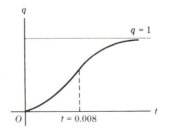

FIG. 5.16

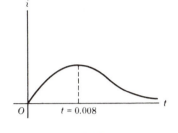

FIG. 5.17

Example. Suppose $L = 0.1$ henry, $R = 10$ ohms, $C = 0.002$ farad, and $E = 50 \sin \omega t$ volts, where ω is a positive constant. Again we suppose that when $t = 0$, $q = 0$ and $i = 0$.

It will be observed in this example that E is a pulsating variable force. Equation (9.2) and its boundary conditions become the differential system

$$(9.2)'' \qquad 0.1 \frac{d^2q}{dt^2} + 10 \frac{dq}{dt} + 500q = 50 \sin \omega t,$$

$$t = 0, \qquad q = 0, \qquad i = 0.$$

To solve this system we note that the corresponding homogeneous equation

$$0.1 \frac{d^2q}{dt^2} + 10 \frac{dq}{st} + 500q = 0$$

has the general solution

$$e^{-50t}(c_1 \cos 50t + c_2 \sin 50t).$$

Accordingly, when $\omega = 50$, the phenomenon of resonance will be encountered (unlike the situation in most mechanical systems, resonance can be a very useful phenomenon—albeit a dangerous one for amateur tinkerers—in electronics).

When $\omega \neq 50$, a particular solution of $(9.2)''$ of the form

$$q = A \sin \omega t + B \cos \omega t \qquad (A, B \text{ constant})$$

can be found. After some computation we find that

$$A = \frac{500(1 - 5000\omega^2)}{(1 - 5000\omega^2)^2 + (100\omega)^2}, \qquad B = -\frac{50{,}000\omega}{(1 - 5000\omega^2)^2 + (100\omega)^2}.$$

The general solution of the differential equation is then

(9.7) $q = A \sin \omega t + B \cos \omega t + e^{-50t}(c_1 \cos 50t + c_2 \sin 50t)$ $(\omega \neq 50)$.

The constants c_1 and c_2 can be determined by the boundary conditions. They are

$$c_1 = -B, \qquad c_2 = \frac{A\omega + 50B}{50}.$$

It will be seen that the contribution to the solution (9.7) of the quantity

(9.8) $e^{-50t}(c_1 \cos 50t + c_2 \sin 50t)$

becomes very small very quickly, as t increases. For this reason the quantity (9.8) is called the *transient* in the system. The quantity

$$A \sin \omega t + B \cos \omega t,$$

on the other hand, is called the *steady-state solution* of the system. The current i is again obtained from the formula

$$i = \frac{dq}{dt}.$$

Two other definitions may be added. The quantity

$$X = L\omega - \frac{1}{C\omega}$$

is called the *reactance* in the circuit, and

$$Z = \sqrt{R^2 + X^2}$$

is called the *impedance*.

Exercises

1. Use equation (9.6) to compute the maximum value of i in the first example.

2. Discuss the simple circuit in which (all quantities are given in standard units)
    (a) $L = 0.01$, $R = 10$, $C = 0.02$, $E = 50$;
    (b) $L = 0.1$, $R = 20$, $C = \frac{1}{360}$, $E = 50 \sin 50t$;
    (c) $L = 0.1$, $R = 10$, $C = 0.002$, $E = 50 \sin 50t$.
    Assume that $q = i = 0$, when $t = 0$.

3. Discuss the simple circuit in which (all quantities are given in standard units)

(a) $L = 0.01,\ R = 20,\ C = 0.02,\ E = 100$;

(b) $L = 0.2,\ R = 10,\ C = \frac{1}{625},\ E = 100 \sin 100t$;

(c) $L = 0.2,\ R = 10,\ C = \frac{1}{625},\ E = 100 \sin 50t$.

Assume that $q = i = 0$, when $t = 0$.

4. Solve the last example in the text when $\omega = 50$. Discuss your answer.

5. Solve Exercise 2 using Laplace transforms.

6. Solve Exercise 3 using Laplace transforms.

Answers

1. $10(\frac{1}{7})^{1/24} = 9.22$ amperes.

2. (a) $q(t) = 1 - \frac{199}{198}e^{-5t} + \frac{1}{198}e^{-995t}$;

(b) $q(t) = -1.141e^{-20t} + 1.635e^{-180t} + 5.43 \sin 50t - 0.494 \cos 50t$

                                              (approximately);

(c) $q(t) = e^{-50t}(\frac{2}{25} \cos 50t + \frac{1}{25} \sin 50t) + \frac{1}{25} \sin 50t - \frac{2}{25} \cos 50t$.

3. (a) $q(t) = -2e^{-2.5t} + 0.0025e^{-2000t} + 2$ (approximately);

(b) $925q(t) = e^{-25t}(32 \cos 50t + 104 \sin 50t) - 44 \sin 100t - 32 \cos 100t$;

(c) $85q(t) = e^{-25t}(16 \cos 50t + 4 \sin 50t) + 4 \sin 50t - 16 \cos 50t$.

6

Elementary oscillation theory

1 Preliminaries

The central ideas in this chapter are of fundamental importance both in the theory and in the applications. They have to do with the behavior of solutions of the second-order linear differential equation

(1.1) $$a(x)y'' + b(x)y' + c(x)y = 0,$$

where $a(x)$, $b(x)$ and $c(x)$ are continuous, with $a(x) > 0$, on an interval I.

If both members of equation (1.1) are multiplied by

$$\frac{1}{a(x)} \, e^{\int_{x_0}^{x} \frac{b(x)}{a(x)} \, dx},$$

where x_0 and x are points of I, equation (1.1) can be rewritten in the so-called *self-adjoint* form

(1.2) $$[r(x)y']' + p(x)y = 0,$$

where $r(x) > 0$, and $r(x)$, $p(x)$ are continuous on I. Here,

$$(1.3) \qquad r(x) = e^{\int_{x_0}^{x} \frac{b(x)}{a(x)} \, dx}, \qquad p(x) = \frac{c(x)}{a(x)} e^{\int_{x_0}^{x} \frac{b(x)}{a(x)} \, dx}.$$

An equation of the form (1.2) can also be put in the form (1.1), provided $r(x)$ is of class C' on I.

We note, for example, that the equations

$$y'' + y = 0, \qquad y'' - y = 0$$

are in self-adjoint form, while the self-adjoint form for the Euler-type equation

$$x^2 y'' - xy' + 2y = 0 \qquad (x > 0)$$

is

$$\left(\frac{1}{x} y' \right)' + \frac{2}{x^3} y = 0.$$

Lemma 1.1. (ABEL'S LEMMA) *If $u(x)$ and $v(x)$ are solutions of equation (1.2), then*

$$(1.4) \qquad r(x)[u(x)v'(x) - v(x)u'(x)] \equiv k,$$

where k is a constant.

The student will observe that the quantity in the brackets in (1.4) is the wronskian of $u(x)$ and $v(x)$.

To prove Abel's lemma we note that because $u(x)$ and $v(x)$ are solutions of (1.2),

$$[r(x)u'(x)]' + p(x)u(x) \equiv 0,$$
$$[r(x)v'(x)]' + p(x)v(x) \equiv 0.$$

If we multiply the first of these identities by $-v(x)$ and the second by $u(x)$ and add, we have

$$(1.5) \qquad u(x)[r(x)v'(x)]' - v(x)[r(x)u'(x)]' \equiv 0.$$

Next, let a be a fixed point and x a variable point of I and integrate both members of (1.5) from a to x:

$$\int_a^x \{u(x)[r(x)v'(x)]' - v(x)[r(x)u'(x)]'\} \, dx \equiv 0.$$

Integration by parts then yields

$$r(x)[u(x)v'(x) - v(x)u'(x)]\Big|_a^x \equiv 0,$$

or

$$r(x)[u(x)v'(x) - v(x)u'(x)] \equiv r(a)[u(a)v'(a) - v(a)u'(a)] = k.$$

The proof of the lemma is complete.

Corollary. *The constant k is zero if and only if the solutions $u(x)$ and $v(x)$ are linearly dependent.*

An immediate consequence of Abel's lemma is the following. Suppose it is desired to determine a second-order linear differential equation (1.2) having two given linearly independent solutions $u(x)$ and $v(x)$. One first calculates the wronskian

$$w(x) = \begin{vmatrix} u(x) & v(x) \\ u'(x) & v'(x) \end{vmatrix}.$$

Abel's formula then yields $r(x)$ immediately inasmuch as

$$r(x) = \frac{k}{w(x)}.$$

Here, k may be taken as any convenient constant $\neq 0$ such that $r(x) > 0$. To calculate $p(x)$ we then employ one of the solutions, say, $u(x)$, to obtain the identity

$$[r(x)u'(x)]' + p(x)u(x) \equiv 0.$$

Let us see how the method works in an example. Suppose the given solutions are $\sin x^2$ and $\cos x^2$, and for simplicity we shall suppose $x > 0$. The wronskian of these two functions is

$$w(x) = \begin{vmatrix} \sin x^2 & \cos x^2 \\ 2x \cos x^2 & -2x \sin x^2 \end{vmatrix} = -2x.$$

Then

$$r(x) = \frac{k}{-2x},$$

or, choosing $k = -1$, we have

$$r(x) = \frac{1}{2x}.$$

To compute $p(x)$ we substitute $\sin x^2$ for y in the equation

$$\left[\frac{1}{2x} y'\right]' + p(x)y = 0,$$

and we have, successively, that $p(x)$ must satisfy the identities

$$(\cos x^2)' + p(x) \sin x^2 \equiv 0,$$
$$-2x \sin x^2 + p(x) \sin x^2 \equiv 0,$$
$$\sin x^2[p(x) - 2x] \equiv 0,$$

and the choice $p(x) = 2x$ becomes apparent. The final equation is then

$$\left[\frac{1}{2x} y'\right]' + 2xy = 0.$$

Exercises

Use the method of the text to determine a second-order linear differential equation, in self-adjoint form, having the given pair of functions as solutions.

1. $\sin x$, $\cos x$.

2. $\sin x^2$, $\cos x^2 - 2 \sin x^2$ $(x > 0)$.

3. x, x^3 $(x > 0)$.

4. $\sin \ln x$, $\cos \ln x$ $(x > 0)$.

5. $\cosh x$, $\sinh x$.

6. x^2, $x^2 \ln x$.

Answers

2. $\left(\frac{1}{2x} y'\right)' + 2xy = 0.$

4. $(xy')' + \frac{1}{x} y = 0.$

6. $(x^{-3}y')' + 4x^{-5}y = 0.$

2 The Sturm separation theorem

We begin with the following result.

Theorem 2.1.   If two solutions $u(x)$ and $v(x)$ of (1.2) [or (1.1)] have a common zero, they are linearly dependent. Conversely, if $u(x)$ and $v(x)$ are linearly dependent solutions, neither identically zero, then if one of them vanishes at $x = x_0$, so does the other.

To prove the first statement of the theorem we employ Abel's formula

$$(2.1) \qquad r(x)[u(x)v'(x) - u'(x)v(x)] \equiv k.$$

Let the common zero of $u(x)$ and $v(x)$ be $x = x_0$, and replace x by x_0 in (2.1). It follows that the constant k is zero, and hence that $u(x)$ and $v(x)$ are linearly dependent.

To prove the converse, suppose $u(x)$ and $v(x)$ are linearly dependent solutions, neither identically zero. Suppose $u(x_0) = 0$. Then, because $k = 0$, we have from (2.1) that

$$r(x_0)[-u'(x_0)v(x_0)] = 0.$$

Since $r(x_0) \neq 0$, it follows that

$$u'(x_0)v(x_0) = 0.$$

But $u'(x_0) \neq 0$ (why?); therefore, $v(x_0) = 0$, and the proof of the theorem is complete.

Theorem 2.2.   Let $u(x)$ and $v(x)$ be differentiable functions on an interval I, and suppose that $u(x)$ has consecutive zeros at $x = x_0$ and $x = x_1$ ($x_0 < x_1$). If the wronskian

$$w(x) = \begin{vmatrix} u(x) & v(x) \\ u'(x) & v'(x) \end{vmatrix} = u(x)v'(x) - u'(x)v(x)$$

is of constant sign on the interval $x_0 \leq x \leq x_1$, $v(x)$ has precisely one zero between x_0 and x_1 (see Fig. 6.1).

The idea of the proof is quite simple. Under the hypotheses $w(x)$ has one sign and is never zero on the interval $x_0 \leq x \leq x_1$. The values of $w(x)$ at $x = x_0$ and $x = x_1$ are compared, and the theorem will follow.

To prove the theorem, then, note first that $v(x_0)$, $v(x_1)$, $u'(x_0)$, and $u'(x_1)$ must all be different from zero. Next,

$$w(x_0) = -u'(x_0)v(x_0),$$
$$w(x_1) = -u'(x_1)v(x_1).$$

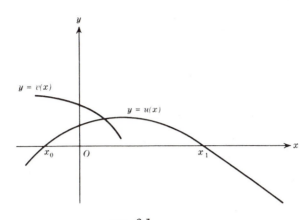

FIG. 6.1

Whether $u(x)$ is always positive or always negative on the interval $x_0 < x < x_1$ (it must be one or the other), the signs of $u'(x_0)$ and $u'(x_1)$ are opposites; that is, if one is positive, the other is negative. Inasmuch as $w(x_0)$ and $w(x_1)$ have the same sign, it follows that $v(x_0)$ and $v(x_1)$ must have opposite signs. But $v(x)$ is a continuous function on the interval $x_0 \le x \le x_1$. Thus, it must vanish at some point between $x = x_0$ and $x = x_1$.

Suppose now that $v(x)$ had a second zero on the interval $x_0 < x < x_1$. By the preceding analysis, reversing the roles of $u(x)$ and $v(x)$, the function $u(x)$ would have at least one zero between those two zeros of $v(x)$, contrary to the hypothesis that x_0 and x_1 are consecutive zeros of $u(x)$.

The proof of the theorem is complete.

Corollary. (THE STURM SEPARATION THEOREM) *If* $u(x)$ *and* $v(x)$ *are linearly independent solutions of* (1.2) [*or* (1.1)], *between two consecutive zeros of* $u(x)$ *there will be precisely one zero of* $v(x)$.

The corollary follows at once from the observation that the wronskian of $u(x)$ and $v(x)$ is never zero on I.

Exercises

1. Put the following differential equations in self-adjoint form (n, α, β constants):

    (a) $x^2 y'' + xy' + (x^2 - n^2)y = 0$ (Bessel equation);
    (b) $(1 - x^2)y'' - 2xy' + n(n + 1)y = 0$ (Legendre equation);

(c) $(1 - x^2)y'' - xy' + n^2y = 0$ (Chebyshev equation);

(d) $xy'' + (1 - x)y' + ny = 0$ (Laguerre equation);

(e) $y'' - 2xy' + 2ny = 0$ (Hermite equation);

(f) $(1 - x^2)y'' + [(\beta - \alpha) - (\alpha + \beta + 2)x]y'$
$$+ n(n + \alpha + \beta + 1)y = 0 \qquad \text{(Jacobi equation).}$$

2. Prove that between every pair of consecutive zeros of $\sin x$ there is one zero of $\sin x + \cos x$. (*Hint.* Use Sturm's theorem.)

3. Prove that between every pair of consecutive zeros of

$$\sin x + \sqrt{2} \cos x$$

there is a zero of $\sin x - \cos x$.

4. Sketch the curves $y = \sin x$ and $y = \sin x - \cos x$ on the interval $-2\pi \le x \le 2\pi$.

5. Show that between every pair of consecutive zeros of $\sin \ln x$ there is a zero of $\cos \ln x$ ($x > 0$).

6. Construct an example of a differential equation in the form (1.1) such that no nonnull solution has more than one zero [thus, there are equations (1.1) to which Sturm's theorem does not apply].

7.* If $u(x)$ is a solution of (1.1) such that $u(x_0) = u(x_1) = 0$ and $u(x) > 0$ ($x_0 < x < x_1$), prove that $u'(x_0) > 0$ and that $u'(x_1) < 0$.

8. Given equation (1.2) with the usual conditions on the coefficients $r(x)$ and $p(x)$, discuss the possibility of solutions $u(x)$ and $v(x)$ having the configuration in Fig. 6.2.

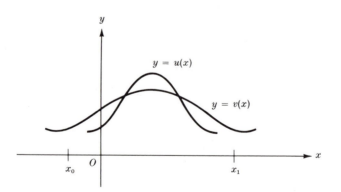

FIG. 6.2

Answers

1.   (a) $(xy')' + \dfrac{x^2 - n^2}{x} y = 0;$

   (b) $[(1 - x^2)y']' + n(n + 1)y = 0;$
   (c) $[(1 - x^2)^{1/2}y']' + n^2(1 - x^2)^{-1/2}y = 0;$
   (d) $(xe^{-x}y')' + ne^{-x}y = 0;$
   (e) $(e^{-x^2}y')' + 2ne^{-x^2}y = 0;$
   (f) $[(1 - x)^{\alpha+1}(1 + x)^{\beta+1}y']' + n(n + \alpha + \beta + 1)(1 - x)^{\alpha}(1 + x)^{\beta}y = 0.$

3   The Sturm comparison theorem

In this section we continue the study of solutions of differential equations of the form

$$(3.1) \qquad\qquad [r(x)y']' + p(x)y = 0,$$

where $r(x) > 0$, and $r(x)$ and $p(x)$ are continuous on the closed interval $a \leq x \leq b$. The student will recall that the Sturm separation theorem asserts that between two consecutive zeros of a solution of (3.1) there appears one zero of every linearly independent solution. Thus, speaking roughly, the number of zeros on an interval of any solution of (3.1) is about the same as the number any other solution possesses.

On the other hand, it is easy to see that solutions (for example, sin 2x) of

$$y'' + 4y = 0$$

oscillate more frequently (that is, have more zeros) on the interval $0 \leq x \leq 2\pi$ than do the solutions of

$$y'' + y = 0$$

on that interval. A typical solution of the last equation is, of course, sin x. The Sturm comparison theorem compares the rates of oscillation of solutions of two equations,

$$(3.2) \qquad\qquad [r(x)y']' + p(x)y = 0,$$

$$(3.3) \qquad\qquad [r(x)z']' + p_1(x)z = 0,$$

where $r(x) > 0$, $r(x)$, $p(x)$, $p_1(x)$ are continuous on $a \leq x \leq b$.

Theorem 3.1. (THE STURM COMPARISON THEOREM) *If a solution $y(x)$ of (3.2) has consecutive zeros at $x = x_0$ and $x = x_1$ $(x_0 < x_1)$, and if $p_1(x) \geq p(x)$ with strict inequality holding for at least one point of the closed interval $[x_0, x_1]$, a solution $z(x)$ of (3.3) which vanishes at $x = x_0$ will vanish again on the interval $x_0 < x < x_1$.*

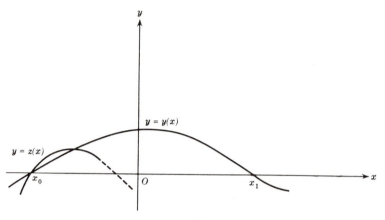

FIG. 6.3

That is to say, speaking roughly, the larger is $p(x)$, the more rapidly the solutions of (3.1) oscillate (see Fig. 6.3).

To prove the theorem, we may suppose without loss in generality that $y(x) > 0$ on the interval $x_0 < x < x_1$, and that $y'(x_0) > 0$, $y'(x_1) < 0$, and $z'(x_0) > 0$. Since $y(x)$ and $z(x)$ are solutions, respectively, of equations (3.2) and (3.3), we have the identities

$$[r(x)y'(x)]' + p(x)y(x) \equiv 0,$$
$$[r(x)z'(x)]' + p_1(x)z(x) \equiv 0.$$

If we multiply the first of these by $-z(x)$ and the second by $y(x)$ and add, we may integrate both members of the resulting identity over the interval $x_0 \leq x \leq x_1$, obtaining, after integration by parts,

$$r(x)[y(x)z'(x) - y'(x)z(x)] \Big|_{x_0}^{x_1} + \int_{x_0}^{x_1} [p_1(x) - p(x)]y(x)z(x)\, dx = 0,$$

or

(3.4) $$r(x_1)y'(x_1)z(x_1) = \int_{x_0}^{x_1} [p_1(x) - p(x)]y(x)z(x)\, dx.$$

Now suppose $z(x) > 0$ on $x_0 < x < x_1$. Then the integral in (3.4) is positive while the left-hand member is not. From this contradiction we infer the truth of the theorem.

Exercises

1.   Solve the differential equation

$$[r(x)y']' = 0$$

and show that no nonnull solution can have more than one zero.

2.   Show that if $p(x) \leq 0$ on I, no nonnull solution of (3.1) can have more than one zero on I. (*Hint.* Compare with solutions of $[r(x)y']' = 0$.)

3.   If nonnull solutions $y(x)$ and $z(x)$, respectively, of the equations

$$x^2 y'' + xy' + (x^2 - 1)y = 0,$$
$$xz'' + z' + xz = 0$$

vanish at $x = 1$, which solution will vanish first after $x = 1$?

4.   Let $y(x)$ and $z(x)$ be nonnull solutions, respectively, of the differential equations

$$y'' + \left[1 - \frac{n^2 - \frac{1}{4}}{x^2}\right]y = 0,$$

$$z'' + \left[1 - \frac{m^2 - \frac{1}{4}}{x^2}\right]z = 0$$

that vanish at $x = 1$. If $m^2 > n^2$, which of the two solutions will vanish first after $x = 1$?

We continue with a very useful result on oscillation of solutions of a differential equation

$$(3.5) \qquad\qquad [r(x)y']' + p(x)y = 0$$

on an interval $I:[a, \infty)$. We suppose, as usual, that $r(x) > 0$ and $r(x)$ and $p(x)$ are continuous on I.

Theorem 3.2. If

$$(3.6) \qquad \int_a^\infty \frac{dx}{r(x)} = +\infty \qquad and \qquad \int_a^\infty p(x)\,dx = +\infty,$$

every solution of equation (3.5) *vanishes infinitely often on* $[a, \infty)$.

To prove the theorem,† suppose that some solution has at most a finite number of zeros on $[a, \infty)$. Then, by the Sturm separation theorem, all (nonnull) solutions have this property and there exists a number $b > a$ such that $y(x) \neq 0$ on $[b, \infty)$. The substitution

$$z = \frac{r(x)y'(x)}{y(x)} \qquad (b \leq x < \infty)$$

---

† A number of proofs of Theorem 3.2 have been given. Some references are: Walter Leighton, "Principal Quadratic Functionals and Self-adjoint Second-order Differential Equations," *Proc. Nat. Acad. of Sciences*, Vol. 35 (1949), p. 193; A. Wintner, "A Criterion of Oscillatory Stability," *Quarterly of Applied Math.*, Vol. 7 (1949), pp. 115–17; Marston Morse, *Variational Analysis: Critical Extremals and Sturmian Extensions*, John Wiley, New York (1973), p. 226. The most elegant proof is surely the proof given above and is due to W. J. Coles, "A Simple Proof of a Well-Known Oscillation Theorem," *Proc. Amer. Math. Soc.*, Vol. 19 (1968), p. 507.

leads to the identity

$$(3.7) \qquad z'(x) + \frac{z^2(x)}{r(x)} + p(x) \equiv 0 \qquad (b \le x < \infty)$$

that $z(x)$ satisfies. Integration of (3.7) from b to x ($> b$) yields

$$(3.8) \qquad z(x) + \int_b^x \frac{z^2(x)}{r(x)} \, dx = z(b) - \int_b^x p(x) \, dx.$$

For x sufficiently large, say, $x \ge c > b$, the right-hand member of (3.8) is negative. Accordingly, $z(x)$ is negative, for $x \ge c$, and

$$(3.9) \qquad z^2(x) > \left[\int_b^x \frac{z^2(x)}{r(x)} \, dx \right]^2.$$

Writing

$$I(x) = \int_b^x \frac{z^2(x)}{r(x)} \, dx,$$

we have from (3.9) that

$$r(x)I'(x) > I^2(x) \qquad (x \ge c).$$

Thus

$$\int_c^x \frac{I'(x)}{I^2(x)} \, dx > \int_c^x \frac{dx}{r(x)},$$

or,

$$\frac{1}{I(c)} > \frac{1}{I(x)} + \int_c^x \frac{dx}{r(x)} \qquad (x > c).$$

This contradicts (3.6), and the proof is complete.

Example. In the equation $y'' + b^2 y = 0$ ($b \ne 0$), $r(x) = 1$, $p(x) = b^2$, conditions (3.6) are satisfied, and all solutions vanish infinitely often on the interval $1 < x < \infty$. This was already known to us, since a solution of the differential equation is $\sin bx$.

Example. In the differential equation

$$(xy')' + \frac{1}{x} y = 0,$$

we note that $r(x) = x$, $p(x) = \frac{1}{x}$. Conditions (3.6) are satisfied on $[1, \infty)$.

Accordingly, all solutions vanish infinitely often on that interval. It is easy to verify that $\sin \ln x$ is a solution of this differential equation on the interval $(0, \infty)$.

Corollary. If $\int_a^\infty \left[xp(x) - \dfrac{1}{4x} \right] dx = +\infty$ $(a > 0)$, *the solutions of the*

differential equation

$$(3.10) \qquad\qquad y'' + p(x)y = 0$$

vanish infinitely often on the interval $[a, \infty)$. *Further, if the above integrand remains less than or equal to zero, for x large, every nonnull solution has at most a finite number of zeros.*

The proof of the corollary involves the substitution

$$(3.11) \qquad\qquad y = x^{1/2}z \qquad (x > 0).$$

Then the differential equation (3.10) becomes

$$(3.12) \qquad\qquad (xz')' + \left[xp(x) - \frac{1}{4x} \right] z = 0.$$

It is clear that if solutions of (3.12) have an infinity of zeros on $[a, \infty)$, so do solutions of (3.10), and conversely. But solutions of (3.12) will vanish infinitely often on $[a, \infty)$ provided

$$\int_a^\infty \frac{dx}{x} = +\infty \qquad and \qquad \int_a^\infty \left[xp(x) - \frac{1}{4x} \right] dx = +\infty.$$

The first condition above is satisfied, and the proof of the first statement in the corollary is complete. Finally, if the second integrand is ≤ 0, for x large, a reference to equation (3.12) indicates that no nonnull solution can vanish infinitely often, by the Sturm comparison theorem. This observation completes the proof of the corollary.

When the function $p(x)$ in equation (3.10) is not eventually positive—for example, if $p(x) = \sin x$—a different kind of oscillation criterion is likely to be required. Such a criterion is provided by the following theorem.

Theorem 3.3. In equation (3.10) let $p(x)$ be of class C'' on $I: [a, \infty)$ and suppose that $|p(x)| \leq m$ and that the functions

$$\int_a^x p(x)\, dx \qquad and \qquad \int_a^x p^3(x)\, dx$$

are bounded on I. If there exists a positive constant α such that

$$(3.13) \quad (\alpha + m)p'(x) + \int_a^x p(x)[p''(x) + 2(\alpha + m)p(x)]\, dx \to +\infty,$$

as $x \to +\infty$, the solutions of the equation (3.10) vanish infinitely often on I.

To prove the theorem set $y = u(x)z$ in (3.10), where

$$u(x) = \alpha + m + p(x).$$

Note that $u(x) \geq \alpha$ on I. Equation (3.10) becomes

(3.14) $(u^2 z')' + u(u'' + pu)z = 0,$

and it is clear that the solutions $y(x)$ of (3.10) will have an infinity of zeros on I if, and only if, the solutions $z(x)$ of (3.14) have this property.

Applying Theorem 3.2 to equation (3.14), we note that solutions of (3.14) will vanish infinitely often on I provided

(3.15) $\int_a^x \frac{dx}{u^2} \to \infty$ and $\int_a^x u(u'' + pu)\, dx \to +\infty,$

as $x \to +\infty$. The first condition in (3.15) is satisfied because

$$u^2(x) \leq (\alpha + 2m)^2.$$

The second integral in (3.15) becomes

$$I_2 = \int_a^x \{(\alpha + m)p'' + p[p'' + (\alpha + m + p)^2]\}\, dx$$

$$= \int_a^x \{(\alpha + m)p'' + p[p'' + 2(\alpha + m)p] + p(\alpha + m)^2 + p^3\}\, dx.$$

The last two terms above are bounded, as $x \to +\infty$, and $I_2 \to +\infty$, if

$$(\alpha + m)p'(x) + \int_a^x p[p'' + 2(\alpha + m)p]\, dx \to +\infty.$$

But this condition is (3.13).

The proof of Theorem 3.3 is complete.

Example.   In the differential equation

(3.16) $y'' + (\sin x)y = 0$ $(I: 0 \leq x < \infty)$

$p(x) = \sin x$, $m = 1$, and the functions

$$\int_a^x \sin x\, dx \qquad \text{and} \qquad \int_a^x \sin^3 x\, dx$$

are bounded on I. We take $u(x) = \alpha + 1 + \sin x$, and observe that

$$\alpha^2 \leq u^2(x) \leq (\alpha + 2)^2.$$

Condition (3.13) becomes

$$\int_a^x \sin x[(2\alpha + 1)\sin x]\, dx \to +\infty,$$

which is clearly satisfied for any choice of $\alpha > 0$.

The solutions of (3.16) must then vanish infinitely often on I.

Exercises

1. Use Theorem 3.2 to show that all solutions of
$$y'' + y = 0$$
    vanish infinitely often on the interval $[0, \infty)$.

2. Do the same as Exercise 1 for Bessel's equation
$$x^2 y'' + xy' + (x^2 - n^2)y = 0 \qquad (1 \le x < \infty).$$

3. Use the corollary to Theorem 3.2 to show that the solutions of the differential equation
$$y'' + kx^\alpha y = 0 \qquad (k > 0, \ \alpha > -2)$$
    are oscillatory on the interval $(0, \infty)$. What happens when $\alpha = -2$?

4. Let $y(x)$ and $z(x)$ be nonnull solutions, respectively, of the differential equations
$$y'' + \left[1 - \frac{n^2 - \frac{1}{4}}{x^2}\right] y = 0,$$
$$z'' + z = 0$$
    that vanish at $x = a > 0$. Which of the two solutions will vanish first after $x = a$ (consider two cases: $n^2 < \frac{1}{4}, n^2 > \frac{1}{4}$).

5. Are the solutions of the differential equation
$$(x + 2)y'' + (x + 1)y = 0$$
    oscillatory on the interval $(0, \infty)$?

6. Prove that the solutions of the differential equation
$$y'' + xy = 0$$
    are oscillatory and that, at least for $x \ge 1$, the distance between consecutive zeros of a solution is less than π.

7.* Prove that the distance between consecutive zeros of a solution of
$$y'' + xy = 0$$
    approaches zero, as $x \to \infty$.

8. Use Theorem 3.3 to show that if m and k are nonzero constants, the solutions of
$$y'' + h(\sin kx)y = 0$$
    vanish infinitely often on $[0, \infty)$.

9. Do the same as in Exercise 8, when $p(x) = h \cos kx$.

10. Show that the solutions of
$$y'' + (\cos^2 x)y = 0$$
    vanish infinitely often on $[0, \infty)$.

4   The conjugate point

Consider the differential equation

$$(4.1) \qquad a(x)y'' + b(x)y' + c(x)y = 0,$$

where $a(x) > 0$ and $a(x)$, $b(x)$, and $c(x)$ are continuous on an interval I of the x-axis. Suppose $y(x) \not\equiv 0$ is a solution such that

$$y(a) = y(b) = 0 \qquad (a, b \in I).\dagger$$

Then the points $x = a$ and $x = b$ of I are said to be *conjugate points* with respect to the differential equation (4.1). Further, if $a < b$ and $y(x) \neq 0$ on the interval (a, b), the point $x = b$ is called the *first conjugate point* of $x = a$.

Recall that every solution $z(x)$ that vanishes at $x = a$ may be written in the form

$$z(x) = ky(x) \qquad (k \text{ constant}).$$

Accordingly, every solution of equation (4.1) that vanishes at $x = a$ also vanishes at $x = b$.

Let $u(x)$ and $v(x)$ be any pair of linearly independent solutions of (4.1). We then have the following lemma.

Lemma.   A necessary and sufficient condition that $x = b$ be conjugate to $x = a$ is that the determinant

$$D = \begin{vmatrix} u(a) & v(a) \\ u(b) & v(b) \end{vmatrix} = 0.$$

The proof of the lemma is easy. Suppose first that $D = 0$. There then exist constants c_1 and c_2, not both zero, such that

$$c_1 u(a) + c_2 v(a) = 0,$$
$$c_1 u(b) + c_2 v(b) = 0.$$

The solution $c_1 u(x) + c_2 v(x)$, with this choice of constants, vanishes at both $x = a$ and $x = b$ and is not identically zero. The proof of the sufficiency is complete.

To prove the necessity of the condition, suppose that $x = b$ is conjugate to $x = a$. There then exists a non-null solution of the differential equation that may be written in the form $c_1 u(x) + c_2 v(x)$ with the property that

$$c_1 u(a) + c_2 v(a) = 0,$$
$$c_1 u(b) + c_2 v(b) = 0.$$

$\dagger$ The notation $(a, b \in I)$ means, of course, that $x = a$ and $x = b$ are points of the interval I.

Inasmuch as c_1 and c_2 are not both zero, it follows that D must be zero, and the proof of the lemma is complete.

A consequence of the lemma is this.

Theorem 4.1.   *If* $x = b$ *is not conjugate to* $x = a$, *the points* (a, c) *and* (b, d) *can be joined by a unique solution curve* $y = y(x)$ *of the differential equation.*

Here c and d are arbitrary numbers. To prove this result it is sufficient to show that there is one and only one pair of constants c_1 and c_2 such that

$$c_1 u(a) + c_2 v(a) = c,$$
$$c_1 u(b) + c_2 v(b) = d.$$

But this follows at once from the fact that $D \neq 0$.

A second illustration of the use of the lemma is the following. Suppose that $q(x) \geq 0$ and that $q(x)$ and $f(x)$ are continuous on an interval $a \leq x \leq b$. To show that a system

$$y'' - q(x)y = f(x),$$
$$y(a) = y(b) = 0$$

always has a unique solution, recall that the general solution of (2) can be written in the form

$$c_1 u(x) + c_2 v(x) + y_0(x),$$

where $y_0(x)$ is any particular solution of (2), and $u(x)$ and $v(x)$ are linearly independent solutions of the corresponding homogeneous differential equation

$$y'' - q(x)y = 0.$$

It is sufficient then to show that there exists a unique pair of constants c_1 and c_2 such that

$$c_1 u(a) + c_2 v(a) + y_0(a) = 0,$$
$$c_1 u(b) + c_2 v(b) + y_0(b) = 0.$$

But this is a consequence of the fact that the determinant

$$D \equiv \begin{vmatrix} u(a) & v(a) \\ u(b) & v(b) \end{vmatrix} \neq 0,$$

for if D were zero, $x = b$ would be conjugate to $x = a$; however, by the Sturm comparison theorem, no non-null solution of the homogeneous differential equation

$$y'' - q(x)y = 0$$

can have more than one zero.

Theorem 4.2.   If $u(x)$ and $v(x)$ are any two linearly independent solutions of (4.1), the zeros $\neq a$ of the solution

$$y(x) = \begin{vmatrix} u(x) & v(x) \\ u(a) & v(a) \end{vmatrix} = v(a)u(x) - u(a)v(x)$$

are the conjugate points of $x = a$.

To prove this theorem, it is sufficient to note that not both $v(a)$ and $u(a)$ are zero inasmuch as $u(x)$ and $v(x)$ are linearly independent. We then have $y(a) = 0$ and $y(x) \not\equiv 0$.

Suppose, for simplicity, that equation (4.1) has the form (or has been put into this form by a transformation)

$$(4.2) \qquad\qquad y'' + p(x)y = 0,$$

where $p(x)$ is positive and continuous on an interval $[0, b]$ and that $y(x) \not\equiv 0$ is a solution such that $y(0) = 0$. It is frequently of importance in the applications to be able to approximate the first conjugate point of $x = 0$, when that conjugate point exists. A useful method that often provides good approximations with very little machinery has been developed† [equations (4.5) and (4.6) that follow].

First, however, we shall provide the general formulas of which (4.5) and (4.6) are a special case.

Let n, the number of divisions of the interval $[0, c]$, be given. We seek to determine a positive number h such that $nh = c_0$, where c_0 is the approximation to c that we seek. Auxiliary variables $c_1, c_2, \ldots, c_n$ $(c_i > 0)$ and $z_1, z_2, \ldots, z_n$ are introduced as follows:

$$c_1^2 = p\left(\frac{h}{2}\right),$$

$$(4.3) \qquad\qquad c_2^2 = p\left(\frac{3h}{2}\right),$$

$$\cdots\cdots\cdots\cdots$$

$$c_n^2 = p\left(\frac{2n-1}{2}h\right);$$

---

† See "On Approximating Conjugate, Focal, and σ-Points for Linear Differential Equations of Second Order," *Annali di Matematica pura ed applicata*, Vol. CVII (1976), pp. 373–81.

$$z_1 = c_1 h,$$

$$\tan z_2 = \frac{c_2}{c_1} \tan z_1,$$

(4.4)
$$\tan z_j = \frac{c_j}{c_{j-1}} \tan (c_{j-1}h + z_{j-1}) \qquad (j = 3, 4, \ldots, n),$$

$$h = \frac{1}{c_n} (\pi - z_n).$$

Equations (4.4) are best solved perhaps, by estimating h, then calculating $z_1, z_2, \ldots, z_n$, successively, from (4.4) and repeating the process using the computed value of h as the next estimate. This is the so-called *method of successive approximations*. The numbers z_i lie on the interval $(0, \pi)$. Difficulties that may arise in this process that require a modification of the method are treated in the reference given in the footnote.

The choice $n = 2$ frequently provides useful results that can be obtained relatively rapidly. In this case, equations (4.3) and (4.4) become, respectively,

$$c_1^2 = p\left(\frac{h}{2}\right),$$
(4.5)
$$c_2^2 = p\left(\frac{3h}{2}\right);$$

$$\tan z_2 = \frac{c_2}{c_1} \tan c_1 h,$$
(4.6)
$$h = \frac{1}{c_2} (\pi - z_2),$$

with $c_0 = 2h$.

It is known, for example, that every nonnull solution of the system

$$y'' + (7 - x^2)y = 0,$$
$$y(0) = 0$$

vanishes at $x = \sqrt{6}/2 = 1.225$. Let us use equations (4.5) and (4.6) to approximate this number. Equations (4.5) become

$$c_1^2 = 7 - \frac{h^2}{4},$$

$$c_2^2 = 7 - \frac{9h^2}{4}.$$

Estimating $h = 0.600$ we calculate c_1 and c_2 and obtain $h = 0.630$ from (4.6). Using the latter value of h yields $h = 0.599$. This suggests using the average as an estimate of h; that is, $h = 0.614$, which in turn yields 0.615. Thus, $c_0 = 2h = 1.229$, which is very close, considering the fact that a very small value of $n (= 2)$ was employed.

Exercises

Use equations (4.5) and (4.6) to approximate the first conjugate point of $x = 0$ for the following differential equations.

1. $y'' + (4 + x^2)y = 0.$

2. $y'' + \dfrac{1}{1 + x} y = 0.$

3. $y'' + (4 + \sqrt{x})y = 0.$

Answers

1.   The precise value of c, to three decimal places, is 1.462.

2.   The precise value of c, to three decimal places, is 5.874.

3.   The precise value of c, to three decimal places, is 1.429.

7

Solutions in power series

It should be regarded as something of a fortunate accident when a differential equation has a simple neat solution in terms of familiar functions. Frequently, however, a series solution is available. In this chapter we shall develop methods of obtaining solutions in the form of *power series* for a number of types of differential equations for which such solutions exist.

1 Power series

An infinite series

$$a_0 + a_1(x - a) + a_2(x - a)^2 + \cdots = \sum_0^\infty a_n(x - a)^n,$$

where $a_0, a_1, a_2, \ldots$ are constants, is a *power series*. Such a series evidently converges when $x = a$. If it converges for any other value of x, it converges on an interval†

$$-b < x - a < b \qquad (0 < b \leq \infty).$$

On this interval it represents a function $f(x)$, the derivative of which may be obtained by term-by-term differentiation; that is,

$$f'(x) = \sum_{0}^{\infty} na_n(x - a)^{n-1} = \sum_{1}^{\infty} na_n(x - a)^{n-1} \qquad (-b < x - a < b).$$

It follows that

$$f''(x) = \sum_{0}^{\infty} n(n - 1)a_n(x - a)^{n-2} = \sum_{2}^{\infty} n(n - 1)a_n(x - a)^{n-2}$$
$$(-b < x - a < b),$$

and so on.

For a large number of the more commonly used power series the *test-ratio test* provides the (open) interval of convergence. Recall that this test states that if the limit

$$\lim_{n \to \infty} \left| \frac{a_{n+1}}{a_n} \right| = k$$

exists and is finite, the (open) interval of convergence of the power series

$$\sum_{0}^{\infty} a_n(x - a)^n$$

is given by (see Fig. 7.1)

$$-\frac{1}{k} < x - a < \frac{1}{k},$$

or

$$a - \frac{1}{k} < x < a + \frac{1}{k}.$$

When $k = 0$, the interval is $-\infty < x < \infty$.

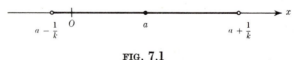

FIG. 7.1

---

† The series may, of course, also converge at one, or both end points of the interval.

For example, in the series

$$\sum_{0}^{\infty} \frac{1}{n+1} (x-2)^n$$

$$a_n = \frac{1}{n+1}, \qquad a_{n+1} = \frac{1}{n+2},$$

and

$$\lim_{n \to \infty} \left| \frac{a_{n+1}}{a_n} \right| = \lim_{n \to \infty} \left| \frac{n+2}{n+1} \right| = 1;$$

accordingly, the (open) interval of convergence of the series is

$$1 < x < 3.$$

(This series also converges when $x = 1$.)

The algebra of power series suggests the following method of solving differential equations. A solution $y(x)$ is assumed in the form

$$y(x) = \sum_{0}^{\infty} a_n(x-a)^n,$$

and after this series is substituted for y in the differential equation, coefficients of like powers of $x - a$ on each side of the equation are equated.

Let us see how the method works in a simple example.

Example. Find a power-series solution of the differential system

$$y'' + y = 0,$$
(1.1)
$$y(0) = 0, \qquad y'(0) = 1.$$

We shall try for a solution in powers of $(x - 0) = x$; that is, we shall attempt to determine constants $a_0, a_1, a_2, \ldots$ such that the power series

$$a_0 + a_1 x + a_2 x^2 + \cdots = \sum_{0}^{\infty} a_n x^n$$

is a solution of (1.1). If we assume that (1.1) has a solution in this form and substitute it for y in the differential equation, we have

$$\sum_{2}^{\infty} n(n-1)a_n x^{n-2} + \sum_{0}^{\infty} a_n x^n \equiv 0,$$

which may be written

$$\sum_{0}^{\infty} (n+2)(n+1)a_{n+2} x^n + \sum_{0}^{\infty} a_n x^n \equiv 0,$$

or

$$\sum_{0}^{\infty} [(n+2)(n+1)a_{n+2} + a_n]x^n \equiv 0.$$

In order that a power series be identically zero, each of its coefficients must be zero; hence,

$$(1.2) \qquad (n + 2)(n + 1)a_{n+2} + a_n = 0 \qquad (n = 0, 1, 2, \ldots).$$

The conditions in (1.1) determine that $a_0 = 0$ and $a_1 = 1$. From (1.2) we then have $0 = a_2 = a_4 = a_6 = \cdots$, and

$$a_{2n+1} = -\frac{a_{2n-1}}{2n(2n + 1)}.$$

Thus,

$$a_3 = -\frac{1}{2 \cdot 3}, \qquad a_5 = \frac{1}{2 \cdot 3 \cdot 4 \cdot 5}, \qquad \cdots,$$

and we have determined the power series

$$(1.3) \qquad y = x - \frac{x^3}{3!} + \frac{x^5}{5!} - \cdots.$$

The test ratio test for this series yields

$$\lim_{n \to \infty} \left| \frac{\dfrac{x^{n+1}}{(n + 1)!}}{\dfrac{x^n}{n!}} \right| = \lim_{n \to \infty} \left| \frac{x}{n + 1} \right| = 0.$$

Thus, this series converges on every finite interval, and consequently provides a solution† of (1.1) for x on the interval $-\infty < x < \infty$. The series (1.3) is seen to be the usual power-series representation of $\sin x$, which was to have been expected.

If $|x|$ is small, the series (1.3) converges very rapidly and lends itself well to computation. Near $x = 100$, however, a great many terms of the series would be required to obtain accuracy to, say, two decimal places. If we were interested, for example, in a series solution $y(x)$ with the property that $y(100) = 2$, $y'(100) = 1$, a more efficient plan would be to attempt to determine the coefficients $a_0, a_1, a_2, \ldots$ so that the power series

$$a_0 + a_1(x - 100) + a_2(x - 100)^2 + \cdots = \sum_0^\infty a_n(x - 100)^n$$

would represent such a solution. The determination of the coefficients is easy in this instance and will be left to the student.

---

† The student will recognize that it is in some sense fortuitous that mathematicians have happened to assign a name ($\sin x$) to the function represented by the power series (1.3). In some respects the series is more useful than the designation "$\sin x$." We can compute from it, for example. We feel more comfortable about the closed expression $\sin x$, partly because it *is* closed, partly because of relative familiarity with the function so designated, and partly because tables of values of the function are available [tables which are computed from expressions like that in (1.3)].

In computing infinite series solutions of differential equations it is helpful to note that

$$\sum_0^\infty b_n x^n = \sum_1^\infty b_{n-1} x^{n-1} = \sum_2^\infty b_{n-2} x^{n-2} = \cdots$$

$$= \sum_{-1}^\infty b_{n+1} x^{n+1} = \sum_{-2}^\infty b_{n+2} x^{n+2} = \cdots.$$

Further,

$$\sum_2^\infty b_n x^n = \sum_1^\infty b_{n+1} x^{n+1} = \sum_0^\infty b_{n+2} x^{n+2} = \cdots.$$

Thus,

$$\sum_p^\infty b_n x^n = \sum_0^\infty b_{n+p} x^{n+p} \qquad (p \text{ any integer}).$$

In deriving equation (1.2) we first had the identity

$$\sum_2^\infty n(n-1) a_n x^{n-2} + \sum_0^\infty a_n x^n \equiv 0.$$

To combine these two series we needed x to appear to the same nominal power in each indicated summation. We elected to rewrite the first series as

$$\sum_0^\infty (n+2)(n+1) a_{n+2} x^n,$$

and equation (1.2) then followed at once.

The next example is somewhat more complicated.

Example. Find a series solution in powers of $(x-1)$ of the differential system

$$(4 - x^2) y'' - 2xy' + 12y = 0,$$

(1.4)

$$y(1) = -7, \qquad y'(1) = 3.$$

It will be convenient first to write the coefficients of the differential equation in powers of $(x-1)$:

$$4 - x^2 = 3 - 2(x-1) - (x-1)^2,$$

(1.5)

$$-2x = -2 - 2(x-1),$$

$$12 = 12.$$

There are a number of ways to obtain equations (1.5). One method is to use Taylor's expansion

$$f(x) = f(1) + \frac{f'(1)}{1!}(x-1) + \frac{f''(1)}{2!}(x-1)^2 + \cdots.$$

When $f(x) = 4 - x^2$, for example, $f(1) = 3$, $f'(1) = -2$, $f''(1) = -2$, and $f^{(n)}(1) = 0$ for $n = 3, 4, \ldots$.

Using equations (1.5) equation (1.4) may then be rewritten in the form

(1.4)' $[3 - 2(x-1) - (x-1)^2]y'' + [-2 - 2(x-1)]y' + 12y = 0.$

Next, we try for a solution of the form

(1.6) $$y = \sum_0^\infty a_n(x-1)^n.$$

Then

$$y' = \sum_0^\infty na_n(x-1)^{n-1},$$

(1.7)

$$y'' = \sum_0^\infty n(n-1)a_n(x-1)^{n-2}.$$

It will simplify computation somewhat if we temporarily set $t = x - 1$ in (1.4)', (1.6), and (1.7). We have, then,

(1.4)'' $(3 - 2t - t^2)y'' + (-2 - 2t)y' + 12y = 0,$

and

(1.6)' $y = \sum_0^\infty a_n t^n,$

$$y' = \sum_0^\infty na_n t^{n-1} = \sum_1^\infty na_n t^{n-1},$$

(1.7)'

$$y'' = \sum_0^\infty n(n-1)a_n t^{n-2} = \sum_1^\infty n(n-1)a_n t^{n-2} = \sum_2^\infty n(n-1)a_n t^{n-2}.$$

Note that

$$(3 - 2t - t^2)y'' = \sum_0^\infty 3n(n-1)a_n t^{n-2} - \sum_0^\infty 2n(n-1)a_n t^{n-1}$$

$$- \sum_0^\infty n(n-1)a_n t^n,$$

(1.8)

$$(-2 - 2t)y' = -\sum_0^\infty 2na_n t^{n-1} - \sum_0^\infty 2na_n t^n,$$

$$12y = \sum_0^\infty 12a_n t^n.$$

Adding the equations in (1.8) yields

$$0 \equiv \sum_0^\infty a_n t^n (12 - n - n^2) - \sum_0^\infty a_n t^{n-1}(2n^2) + \sum_0^\infty a_n t^{n-2}(3n^2 - 3n).$$

This identity can be written in the form

$$\sum_0^\infty a_n(12 - n - n^2)t^n - \sum_{-1}^\infty 2a_{n+1}(n+1)^2 t^n$$
$$+ \sum_{-2}^\infty 3a_{n+2}(n^2 + 3n + 2)t^n \equiv 0,$$

or

(1.9) $\quad \sum_0^\infty a_n(12 - n - n^2)t^n - \sum_0^\infty 2a_{n+1}(n+1)^2 t^n$
$$+ \sum_0^\infty 3a_{n+2}(n^2 + 3n + 2)t^n \equiv 0.$$

The three series in (1.9) are all nominally series in t^n, and their first terms are provided by $n = 0$; accordingly,

$$3a_{n+2}(n+1)(n+2) - 2a_{n+1}(n+1)^2 + a_n(4+n)(3-n) = 0$$
$$(n = 0, 1, 2, \ldots),$$

or

(1.10) $\quad a_{n+2} = \dfrac{2(n+1)^2 a_{n+1} + (n-3)(n+4)a_n}{3(n+1)(n+2)} \qquad (n = 0, 1, 2, \ldots).$

From the conditions in (1.4) we have

$$a_0 = -7,$$
$$a_1 = 3.$$

It follows from (1.10), upon setting $n = 0, 1, 2, 3$, successively, that

$$a_2 = 15,$$
$$a_3 = 5,$$
$$a_4 = 0,$$
$$a_5 = 0;$$

consequently, $a_n = 0$ $(n = 4, 5, \ldots)$. Accordingly, the series terminates, and the solution we seek is the polynomial

(1.5)' $\qquad -7 + 3(x - 1) + 15(x - 1)^2 + 5(x - 1)^3.$

We note that written in powers of x the solution (1.5)' becomes $-12x + 5x^3$.

An alternate method for solving (1.4) in powers of $(x - 1)$ would be to set $t = x - 1$ at the outset and note that

$$\frac{dy}{dx} = \frac{dy}{dt}, \qquad \frac{d^2y}{dx^2} = \frac{d^2y}{dt^2}.$$

The system (1.4) would then become

(1.4)'''
$$(3 - 2t - t^2)\ddot{y} - 2(1 + t)\dot{y} + 12y = 0,$$
$$y(0) = -7, \qquad y'(0) = 3.$$

We would then try for a solution of the form

$$y = \sum_0^\infty a_n t^n$$

obtaining eventually

$$y = -7 + 3t + 15t^2 + 5t^3$$

as a solution of (1.4)'''. Setting $t = x - 1$ in this result then yields (1.5)'.

In the next section we shall see still another way to solve such a system as (1.4).

Exercises

Find power-series solutions of the following differential systems.

1. $y'' + y = 0, y(0) = 1, y'(0) = 0$ (in powers of x).

2. $y'' + y = 0, y(0) = 1, y'(0) = 1$ (in powers of x).

3. $y'' - 3y' + 2y = 0, y(0) = 0, y'(0) = 1$ (in powers of x).

4. $y'' + 4y' + 4y = 0, y(0) = 0, y'(0) = 0$ (in powers of x).

5. $(1 - x^2)y'' - 2xy' + 6y = 0, y(0) = -1, y'(0) = 0$ (in powers of x).

6. $y'' + y = 0, y(10) = 0, y'(10) = 1$ (in powers of $x - 10$).

7. $y'' - 3y' + 2y = 0, y(-4) = 0, y'(-4) = 1$ (in powers of $x + 4$).

8. $x^2y'' - xy' + 2y = 0, y(1) = 0, y'(1) = 1$ (in powers of $x - 1$).

9. $x^2y'' + xy' - 4y = 0, y(1) = 0, y'(1) = 1$ (in powers of $x - 1$).

10. $x^2y'' - 2xy' + 2y = 0, y(-2) = 1, y'(-2) = 0$ (in powers of $x + 2$).

11. $y'' - 2xy' + 6y = 0, \qquad y(0) = 0, \qquad y'(0) = -12.$

12. $y'' - 2xy' + 8y = 0, \qquad y(0) = 3, \qquad y'(0) = 0.$

Answers

1. $1 - (x^2/2!) + (x^4/4!) - \cdots$.

3. $x + (3x^2/2!) + (7x^3/3!) + (15x^4/4!) + \cdots$.

5. $-1 + 3x^2$.

6. $(x - 10) - (1/3!)(x - 10)^3 + (1/5!)(x - 10)^5 - \cdots$.

7. $(x + 4) + \dfrac{3(x + 4)^2}{2!} + \dfrac{7(x + 4)^3}{3!} + \dfrac{15(x + 4)^4}{4!} + \cdots$.

8. $(x - 1) + [(x - 1)^2/2!] - [2(x - 1)^3/3!] + [4(x - 1)^4/4!] - [10(x - 1)^5/5!] + \cdots$.

9. $\frac{1}{4}[4(x - 1) - 2(x - 1)^2 + 4(x - 1)^3 - 5(x - 1)^4 + 6(x - 1)^5 - \cdots]$.

11. $8x^3 - 12x$.

2 Alternate method

For many problems a method other than that given in Section 1 may prove to be simpler.

Example. Consider the differential system

(2.1)
$$y'' + y = 2x - 1,$$
$$y(1) = 1, \qquad y'(1) = 3.$$

Suppose there is a solution $y(x)$ of (2.1) which can be represented as a power series in $(x - 1)$. Then by Taylor's theorem

(2.2) $$y(x) = y(1) + \frac{y'(1)}{1!}(x - 1) + \frac{y''(1)}{2!}(x - 1)^2 + \cdots.$$

We shall, accordingly, have the series we seek if we can evaluate the constants

(2.3) $$y(1), y'(1), y''(1), \cdots.$$

The first two terms of this sequence are given by the side conditions in (2.1). From the differential equation we then have that

(2.4) $$y''(x) = -y(x) + 2x - 1.$$

It follows that

$$y''(1) = -y(1) + 1$$
$$= 0.$$

Next, we differentiate both members of (2.4), obtaining

(2.5) $y'''(x) = -y'(x) + 2;$

hence,

$$y'''(1) = -1.$$

We continue the process, differentiating (2.5). This yields

$$y''''(1) = -y''(1) = 0.$$

In this fashion each member of (2.3) may be determined successively, and we have

(2.6) $y(x) = 1 + \dfrac{3}{1!}(x-1) - \dfrac{1}{3!}(x-1)^3 + \dfrac{1}{5!}(x-1)^5 - \cdots.$

In this case, if we note that we may write the general solution of the equation $y'' + y = 0$ in the form

$$c_1 \sin(x-1) + c_2 \cos(x-1),$$

the system (2.1) can be solved by inspection. We have

(2.7) $y = 2x - 1 + \sin(x-1).$

Clearly, (2.6) is the formal expansion of (2.7) in powers of $(x-1)$.

This alternate method is likely to be simpler to apply than the method given in Section 1 when the differential equation is nonlinear.

Example.   Consider the differential system,

(2.8)
$$y' = 1 + y^2,$$
$$y(0) = 0.$$

The solution of this system is clearly $\tan x \left(-\dfrac{\pi}{2} < x < \dfrac{\pi}{2} \right)$. Suppose we try to determine the solution $y(x)$ in the form

$$y(x) = y(0) + \frac{y'(0)}{1!}x + \frac{y''(0)}{2!}x^2 + \cdots.$$

If $y(x)$ is assumed to be a solution of (2.8), we have at once

(2.9) $y'(x) = 1 + y^2(x),$
$$y(0) = 0,$$
$$y'(0) = 1 + y^2(0) = 1.$$

We differentiate (2.9), obtaining

(2.10) $y''(x) = 2y(x)\,y'(x).$

It follows that
$$y''(0) = 0.$$
Similarly,
$$y'''(x) = 2y(x)y''(x) + 2y'^2(x);$$
thus,
$$y'''(0) = 2.$$

The process may be continued indefinitely, and it will yield as many terms of the series as patience and computing facilities permit. We have, after two more differentiations,

$$y(x) = x + \frac{x^3}{3} + \frac{2x^5}{15} + \cdots.$$

The method of Section 1 is more complicated in that it involves squaring a power series. Using that method we would suppose

$$y(x) = \sum_1^\infty a_n x^n,$$

$$y'(x) = \sum_1^\infty na_n x^{n-1},$$

since a_0 is seen to be zero from (2.8). Then, formally,

$$y^2(x) = a_1^2 x^2 + 2a_1 a_2 x^3 + (a_2^2 + 2a_1 a_3)x^4 + 2(a_1 a_4 + a_2 a_3)x^5 + \cdots.$$

Setting $y'(x) = 1 + y^2(x)$ yields

(2.11) $a_1 + 2a_2 x + 3a_3 x^2 + 4a_4 x^3 + 5a_5 x^4 + \cdots$
$$= 1 + a_1^2 x^2 + 2a_1 a_2 x^3 + (a_2^2 + 2a_1 a_3)x^4 + \cdots.$$

From (2.11) we have, successively,

$$a_1 = 1, \qquad a_2 = 0, \qquad a_3 = \tfrac{1}{3}, \qquad a_4 = 0, \qquad a_5 = \tfrac{2}{15}, \cdots,$$

which agrees with our previous result.

Critique.   The general problem of convergence of power series to functions which represent solutions of the differential systems belongs in the realm of the theory of functions of a complex variable. In the examples and exercises in this chapter we are dealing with differential equations of the form

(2.12) $y^{(n)} = f[x, y, y', \ldots, y^{(n-1)}],$

where the right-hand member of (2.12) is, in general, a very well behaved (analytic) function of the $n + 1$ variables indicated. For example, the differential equation

$$(2.13) \qquad x^2 y'' - 2xy' - 4y = 0,$$

may be written in the form

$$y'' = \frac{2}{x} y' + \frac{4}{x^2} y.$$

Here,

$$f(x, y, y') = \frac{2}{x} y' + \frac{4}{x^2} y,$$

and this function, regarded for the moment as a function of the three independent variables x, y, and y', is well behaved provided $x \neq 0$.

The general situation will be clear from the differential equation $y'' = f(x, y, y')$. If $f(x, y, y')$ is well behaved (analytic) in a region containing the point $x = a$, $y = b$, $y' = c$, there then exists a convergent power-series solution $\sum a_n(x - a)^n$ of the differential system

$$y'' = f(x, y, y'),$$

$$y(a) = b, \qquad y'(a) = c,$$

which is valid on an interval $-k < x - a < k$, where k is some number such that $0 < k \leq \infty$.

In particular, the differential system

$$y'' = \frac{2}{x} y' + \frac{4}{x^2} y,$$

$$(2.13)'$$

$$y(1) = 0, \qquad y'(1) = 1$$

will then possess a convergent power-series solution of the form

$$y(x) = \sum_0^\infty a_n(x - 1)^n$$

for values of x on an interval $-k < x - 1 < k$, where $k > 0$. (In this particular case it can be shown that $k = 1$.)

Again, consider the differential system

$$y'' = -\frac{3y'^2}{y},$$

$$(2.14)$$

$$y(0) = 1, \qquad y'(0) = \tfrac{1}{4}.$$

The right-hand member of the differential equation in (2.14) is well behaved near the point $x = 0$, $y = 1$, $y' = \tfrac{1}{4}$. Accordingly, there exists a convergent power-series solution of the differential system (2.14) of the form

$$y(x) = \sum_0^\infty a_n x^n$$

for x on some interval $-k < x < k$ $(0 < k \leq \infty)$.

Exercises

1. Find a power-series solution of the differential system
$$yy'' + 3y'^2 = 0,$$
$$y(0) = 1, \qquad y'(0) = \tfrac{1}{4}.$$

2. Find a power-series solution in powers of $(x - 1)$ of the differential system
$$x^2y'' - 2xy' + 2y = 2,$$
$$y(1) = 1, \qquad y'(1) = 1.$$

3. Find a power-series solution of the differential system
$$y'' + 4y' + 3y = 0,$$
$$y(0) = 1, \qquad y'(0) = -1.$$

4. Find a power-series solution of the differential system
$$2yy'' + y'^2 - 3y^2 = 0,$$
$$y(0) = 1, \qquad y'(0) = 1.$$
Identify the solution.

5. Find a power-series solution in powers of $(x - 1)$ of the system
$$x^2y'' - 2xy' + 2y = 2 - 2x^3,$$
$$y(1) = 0, \qquad y'(1) = 0.$$

6. Find a power-series solution in powers of x of the system
$$y'' = 1 + y'^2,$$
$$y(0) = 1, \qquad y'(0) = 1.$$

7. Find a power-series solution in powers of x of the system
$$2yy'' = 1 + y'^2,$$
$$y(0) = 2, \qquad y'(0) = -1.$$

8. Find a power-series solution in powers of x of the system
$$y''' - y'' + y' - y = x - 2,$$
$$y(0) = 1, \qquad y'(0) = 3, \qquad y''(0) = 4.$$

9. Find a power-series solution in powers of x of the system
$$y^{IV} - y = 3 - 2x + x^2,$$
$$y(0) = -2, \qquad y'(0) = 3,$$
$$y''(0) = 1, \qquad y'''(0) = 1.$$

10. Find a power-series solution of the differential system
$$y'''' - 2y''' + 3y'' - 2y' + 2y = 0,$$
$$y(0) = 0, \qquad y'(0) = 1, \qquad y''(0) = 0, \qquad y'''(0) = -1.$$
Identify the solution, and thus factor the characteristic polynomial
$$m^4 - 2m^3 + 3m^2 - 2m + 2.$$

Answers

1. $1 + \dfrac{x}{2^2} - \dfrac{3x^2}{2^5} + \dfrac{7x^3}{2^7} - \dfrac{77x^4}{2^{11}} + \cdots$.

2. $1 + (x - 1) + (x - 1)^2$.

5. $1 - 3x + 3x^2 - x^3$.

6. $1 + x + x^2 + \frac{2}{3}x^3 + \frac{1}{3}x^4 + \frac{2}{3}x^5 + \frac{32}{45}x^6 + \cdots$.

8. $1 + 3x + 2x^2 - \dfrac{2x^3}{3!} + \dfrac{2x^5}{5!} + \dfrac{4x^6}{6!} + \dfrac{4x^8}{8!} - \cdots$.

9. $-2 + 3x + \frac{1}{2}x^2 + \dfrac{x^3}{3!} + \dfrac{x^4}{4!} + \dfrac{x^5}{5!} + \dfrac{3x^6}{6!} + \dfrac{x^7}{7!} + \cdots$.

3   Regular singular points

In this section we shall confine our attention to homogeneous linear differential equations of second order; that is, equations of the form

$$a(x)y'' + b(x)y' + c(x)y = 0,$$

where $a(x)$, $b(x)$, and $c(x)$ are well behaved (analytic) on an interval I. The function $a(x)$ may vanish at isolated points of I; otherwise, it will be different from zero. The points of I where $a(x) \neq 0$ are called *regular* points of the differential equation, while a point at which $a(x) = 0$ is called a *singular* point. A singular point $x = a$ will be called a *regular singular* point if the differential equations of second order—that is, equations of the form

(3.1) $(x - a)^2 p_0(x)y'' + (x - a)p_1(x)y' + p_2(x)y = 0,$

where $p_0(x)$, $p_1(x)$, and $p_2(x)$ are representable by power series in powers of $(x - a)$, which are valid in a common interval $|x - a| < h$ $(h > 0)$, and where $p_0(x) \neq 0$ in this interval. We shall suppose also that a is real and that the functions $p_i(x)$ are real for real values of x.

Many of the classical differential equations of interest to the physicist as well as to the mathematician may be written in this form. Perhaps the two best-known examples are the following:

(3.2)
$$x^2 y'' + xy' + (x^2 - n^2)y = 0 \qquad \text{(Bessel)},$$
$$(1 - x^2)y'' - 2xy' + n(n + 1)y = 0 \qquad \text{(Legendre)}.$$

The former has a regular singular point at $x = 0$, and the latter has a regular singular point at $x = 1$ and at $x = -1$.

The definition of a regular singular point as formulated above includes as a special case a regular (nonsingular) point. Thus, the differential equation

$$y'' + y = 0$$

has a regular singular point at $x = 1$, for example, for we may write the differential equation in the form

$$(x - 1)^2 y'' + (x - 1)^2 y = 0,$$

where $p_0(x) = 1$, $p_1(x) = 0$, and $p_2(x) = (x - 1)^2$. Accordingly, the basic theorems we state concerning solutions of equations (3.1) in the neighborhood of a regular singular point are valid also near a regular point of the differential equation.

In this section, however, we shall be concerned with regular singular points that are, in fact, also singular points of the differential equations.

Perhaps the simplest differential equation with a regular singular point is the Euler-type equation (Chapter 3, Section 5)

$$(3.3) \qquad\qquad ax^2 y'' + bxy' + cy = 0 \qquad (a \neq 0),$$

where a, b, and c are constants.

Experience with differential equations of the form (3.3) suggests that we try to determine a power series $P(x)$, valid for x near a, such that

$$(3.4) \qquad\qquad y = (x - a)^r P(x)$$

is a solution of the differential equation. Substituting (3.3) in (3.1) leads to the identity

$$(3.5) \qquad (x - a)^2 p_0 P'' + (x - a)(2r p_0 + p_1)P' \\ + [p_0 r(r - 1) + r p_1 + p_2]P \equiv 0.$$

In particular, if there is a solution of the type we seek, this identity must hold when $x = a$. Thus, if (3.4) provides a solution of equation (3.1), r must be a root of the quadratic equation

$$(3.6) \qquad\qquad p_0(a) r(r - 1) + p_1(a) r + p_2(a) = 0.$$

Equation (3.6) is known as the *indicial* equation associated with the regular singular point $x = a$ of equation (3.1), and its roots are called the *indicial roots*.

Equation (3.6) is a quadratic, and its roots may be real and unequal, real and equal, or conjugate imaginaries. We shall limit ourselves to a discussion of differential equations for which the roots r_1 and r_2 of (3.6) are real, and we shall always suppose that $r_1 \geq r_2$.

A difficulty arises, however, even when the roots are real; suppose, for example, $r_1 = \sqrt{2}$, $r_2 = -\sqrt{2}$. Then for $x < a$, the quantity $(x - a)^r$ requires attention. In this case, we would consider a solution

(3.7) $$y = (-x + a)^r P(x).$$

Both (3.4) and (3.7) are included, however, in

$$y = |x - a|^r P(x),$$

and (3.5) and (3.6) will follow as written above.

The following result is fundamental (and is valid even if the indicial roots are imaginary numbers).

Theorem 3.1. *If the roots r_1 and r_2 of the indicial equation differ by a number which is not an integer, there exist two linearly independent solutions of* (3.1),

$$|x - a|^{r_1} P_1(x) \qquad and \qquad |x - a|^{r_2} P_2(x),$$

where $P_1(x)$ and $P_2(x)$ are representable as power series in $(x - a)$ which are valid in the interval $|x - a| < h$ and $P_1(a) \neq 0$, $P_2(a) \neq 0$.

If the difference $r_1 - r_2$ is a positive integer or zero, there exist linearly independent solutions of (3.1),

$$y_1(x) = |x - a|^{r_1} P_1(x),$$
$$y_2(x) = |x - a|^{r_2} P_2(x) + C y_1(x) \ln |x - a|,$$

where the functions $P_1(x)$ and $P_2(x)$ are representable by power series in $(x - a)$ which are valid in the interval $|x - a| < h$ and $P_1(a) \neq 0$, $P_2(a) \neq 0$. The constant C is not zero when $r_1 = r_2$; if $r_1 - r_2$ is a positive integer, C may or may not be zero.

The proof of the theorem is omitted.†

We shall turn our attention to solving equations (3.1) when $a = 0$. Equation (3.1) then becomes

(3.1)′ $$x^2 p_0(x) y'' + x p_1(x) y' + p_2(x) y = 0.$$

---

† The proof of the theorem (due to Fuchs) from which Theorem 3.1 follows is to be found, for example, in Brauer and Nohel, *Ordinary Differential Equations*, Ch. 4, W. A. Benjamin, New York (1967).

The existence proof of the original theorem amounts first to demonstrating the existence of a convergent power series solution $P(x)$ of (3.5) corresponding to the index r_1 which has the greater real part. Obtaining in this fashion the solution

$$y_1(x) = (x - a)^{r_1} P(x),$$

one makes the substitution in (3.1) of $y = (x - a)^{r_1} P(x) w$ to determine w and hence a second linearly independent solution.

It should be pointed out that the functions $P_1(x)$ and $P_2(x)$ may be complex functions when the indicial roots r_1 and r_2 are conjugate complex numbers.

The student will recognize that then only obvious changes are required for the case $a \neq 0$.

Theorem 3.1 insures that there always exists a solution of $(3.1)'$ of the form

$$y_1(x) = |x|^{r_1} \sum_0^\infty c_n x^n \qquad (c_0 \neq 0),$$

where r_1 is the larger of the two roots $(r_1 \geq r_2)$, and the series $\sum c_n x^n$ is valid in some interval about $x = 0$. To determine this solution then, we assume for the moment that $x > 0$ and substitute

$$(3.8) \qquad\qquad y = x^{r_1} \sum_0^\infty c_n x^n \qquad (c_0 \neq 0),$$

in equation $(3.1)'$ and determine the coefficients $c_0, c_1, c_2, \ldots$ of the power series. We may always assume, if we choose, that $c_0 = 1$. After determining $c_0, c_1, c_2, \ldots$ we write the solution (3.8) in the form

$$y_1(x) = |x|^{r_1} \sum_0^\infty c_n x^n,$$

unless r_1 is an integer (when this refinement is unnecessary).

The solution $y_1(x)$ corresponding to the larger of the two roots of the indicial equation is called a *principal solution* associated with the point $x = 0$. There are several methods available for computing a solution linearly independent of $y_1(x)$. In any, the computation frequently becomes rather cumbersome. The classical method is to substitute $y = y_1(x)z$ in (3.1), obtaining

$$(3.9) \qquad x p_0(x) y_1(x) z'' + [2x p_0(x) y_1'(x) + p_1(x) y_1(x)] z' = 0.$$

This is a linear homogeneous differential equation of the first order in z' which may be written as

$$(3.9)' \qquad\qquad z'' + \left[2 \frac{y_1'(x)}{y_1(x)} + \frac{p_1(x)}{x p_0(x)} \right] z' = 0.$$

Its solution is

$$(3.10) \qquad\qquad w = z' = c \frac{1}{y_1^2(x)} e^{-\int \frac{p_1(x)\,dx}{x p_0(x)}}.$$

The operations indicated in (3.10) may be carried out in terms of power series. From the result we may compute z as a power series. Finally, the product $y_1(x)z$ yields the desired solution.

When the difference $r_1 - r_2$ is not an integer, it is, of course, simpler to substitute

(3.11) $$y = x^{r_2} \sum_0^\infty b_n x^n \qquad (b_0 \neq 0)$$

in the differential equation (3.1)' and determine the coefficients b_n directly.

A more promising approach to the second solution is the following. When $r_1 - r_2$ is a positive integer, the constant C in Theorem 3.1 *may* be zero; thus, as a procedural matter, one may try for a solution of (3.1)' of the form (3.11). If such a (nonnull) solution can be determined, the required second solution has been found. If the second solution does not have the form (3.11), as is the case when $r_1 - r_2 = 0$, we observe from Theorem 3.1 that a second solution has the form (for $x > 0$)

(3.12) $$y_2(x) = x^{r_2} \sum_0^\infty b_n x^n + y_1(x) \ln x,$$

which we write in the form

(3.13) $$y_2 = z + y_1 \ln x.$$

Substituting y_2 from (3.13) for y in (3.1)' we have

$$\ln x [x^2 p_0 y_1'' + x p_1 y_1' + p_2 y_1] + [x^2 p_0 z'' + x p_1 z' + p_2 z]$$
$$+ [2x p_0 y_1' + (p_1 - p_0) y_1] = 0.$$

Inasmuch as y_1 is a solution of (3.1)', the above equation can be written

(3.14) $$x^2 p_0 z'' + x p_1 z' + p_2 z + [2x p_0 y_1' + (p_1 - p_0) y_1] = 0,$$

where

$$z = x^{r_2} \sum_0^\infty b_n x^n.$$

It follows from (3.14) that when the quantity

(3.15) $$2x p_0(x) y_1'(x) + [p_1(x) - p_0(x)] y_1(x) \equiv 0,$$

$z(x)$ may be taken as $y_1(x)$, and the second solution becomes

$$y_2(x) = y_1(x)[1 + \ln x].$$

But $y_2(x) - y_1(x)$ is also a solution of the differential equation, so two linearly independent solutions are

$$y_1(x), \qquad y_1(x) \ln x.$$

Conversely, if $y_1(x)$ and $y_1(x) \ln x$ $(x > 0)$ are solutions of a differential equation (3.1)', where $y_1(x) \not\equiv 0$ is a function representable† as a power

---

† It is sufficient that $y_1(x)$ be of class C' and $\neq 0$ on an interval containing the origin.

series near $x = 0$, it is not difficult to calculate that $y_1(x)$ and $y_1(x) \ln x$ are linearly independent solutions of the differential equation

$$x^2 u^2 y'' + x(u^2 - 2xuu')y' + x(2xu'^2 - xuu'' - uu')y = 0.$$

Here, $p_1 = u^2$, $p_0 = u^2 - 2xuu'$, and (3.15) becomes

$$2xu^2 u' + [(u^2 - 2xuu') - u^2]u \equiv 0.$$

It can be shown that condition (3.15) is equivalent to the condition

(3.15)′
$$2x \left[\frac{p_1(x)}{p_0(x)} \right]' \equiv 4 \left[\frac{p_2(x)}{p_0(x)} \right] - \left[\frac{p_1(x) - p_0(x)}{p_0(x)} \right]^2.$$

We have then the following result.

Theorem 3.2. *A necessary and sufficient condition that the differential equation (3.1)′ have linearly independent solutions*

$$y_1(x), \qquad y_1(x) \ln |x| \qquad (y_1(x) \not\equiv 0),$$

is that (3.15)′ hold.

Let us apply the method to the equation

(3.16)
$$x^2 y'' + xy' + (x - 1)y = 0 \qquad (x > 0).$$

The indicial roots associated with $x = 0$ are $r_1 = 1$, $r_2 = -1$, and suppose it has been verified (see Exercise 17) that a principal solution is

(3.17)
$$y_1(x) = x \sum_0^\infty c_n x^n,$$

where

(3.18)
$$c_0 = 1, \qquad c_n = (-1)^n \frac{2}{n!(n + 2)!} \qquad (n = 1, 2, \dots).$$

The substitution in (3.16) of

$$y = x^{-1} \sum_0^\infty b_n x^n \qquad (b_0 \neq 0)$$

will not lead to a solution, and let us assume that this too has been discovered. Next, we observe that

$$2xp_0 y_1' + (p_1 - p_0)y_1 = \sum_0^\infty 2(n + 1)c_n x^{n+1},$$

and the equation (3.14) becomes

$$x^2 z'' + xz' + (x - 1)z + \sum_0^\infty 2(n + 1)c_n x^{n+1} = 0.$$

When we substitute

$$z = x^{-1} \sum_0^\infty b_n x^n \qquad (b_0 \neq 0)$$

in the last equation, we first have

$$\sum_0^\infty b_n(n^2 - 2n)x^{n-1} + \sum_0^\infty b_n x^n + \sum_0^\infty 2(n + 1)c_n x^{n+1} = 0.$$

This equation can be written, after the usual adjustments, as

$$b_0 - b_1 + (b_1 + 2)x + \sum_2^\infty [(n^2 - 1)b_{n+1} + b_n + 2nc_{n-1}]x^n = 0.$$

It follows that

$$(3.19) \quad b_1 = b_0 = -2, \qquad (n^2 - 1)b_{n+1} = -b_n - 2nc_{n-1} \qquad (n = 2, 3, \ldots),$$

and, hence that

$$(3.20) \quad b_{n+1} = \frac{-b_n}{n^2 - 1} + (-1)^n \frac{4n}{(n^2 - 1)(n - 1)! \, (n + 1)!}$$

$$(n = 2, 3, \ldots),$$

where b_2 is arbitrary. Accordingly, a second linearly independent solution of equation (3.16) is

$$y_1(x) \ln |x| + \sum_0^\infty b_n x^n,$$

where $y_1(x)$ is given by (3.17) and (3.18) and the coefficients b_n are given by (3.19) and (3.20).

Exercises

Find the general solutions of the following differential equations.

1. $x^2 y'' - x(1 + 2x)y' + (1 + x + x^2)y = 0.$

2. $x^2 y'' + x(3 - 2x)y' + (1 - 3x + x^2)y = 0.$

3. $4x^2 y'' + 8x^2 y' + (1 + 4x^2)y = 0.$

4. $x^2 y'' - 2xy' + (2 + x^2)y = 0.$

5. $xy'' - y' + 4x^3 y = 0.$

6. $(x - 2)^2 y'' - 4(x - 2)y' + (x^2 - 4x + 10)y = 0.$

7. $(x - 1)^2 y'' + (x - 1)(1 - 2x)y' + (1 - x + x^2)y = 0.$

8. $xy'' + (1 - 2x)y' + (x - 1)y = 0.$

9. $x^2y'' + x(1 - 3x)y' + (2x^2 - 3x - 4)y = 0.$

10. $x^2y'' + (5x - 2x^2)y' + (x^2 - 5x + 4)y = 0.$

11. $x^2y'' - 3x^2y' + (2x^2 + \frac{1}{2}x + \frac{1}{4})y = 0.$

12. $x^2y'' - 2\alpha xy' + [\alpha(\alpha + 1) + x^2]y = 0$ (α constant).

13. Find the general solution of the differential equation
$$x^2y'' + xy' + (x^2 - \tfrac{1}{4})y = 0.$$

14. Find the solution of the system
$$xy'' + (1 - x)y' + 3y = 0$$
$$y(0) = 6, \qquad y'(0) = -18.$$

15. Find the general solution of the differential equation
$$xy'' - 2y' + 9x^5y = 0.$$

16. Find the general solution of Bessel's equation
$$x^2y'' + xy' + (x^2 - n^2)y = 0,$$
when $n = 0.$

17. Verify the formula for c_n in (3.17).

Answers

1. $\left(x + \dfrac{x^2}{1!} + \dfrac{x^3}{2!} + \dfrac{x^4}{3!} + \cdots\right)(c_1 + c_2 \ln |x|).$

2. $x^{-1}\left(1 + \dfrac{x}{1!} + \dfrac{x^2}{2!} + \dfrac{x^3}{3!} + \cdots\right)(c_1 + c_2 \ln |x|).$

3. $|x|^{1/2}\left(1 - \dfrac{x}{1!} + \dfrac{x^2}{2!} - \dfrac{x^3}{3!} + \cdots\right)(c_1 + c_2 \ln |x|).$

4. $c_1\left(x^2 - \dfrac{x^4}{3!} + \dfrac{x^6}{5!} - \cdots\right) + c_2\left(x - \dfrac{x^3}{2!} + \dfrac{x^5}{4!} - \cdots\right).$

5. $c_1\left(x^2 - \dfrac{x^6}{3!} + \dfrac{x^{10}}{5!} - \cdots\right) + c_2\left(1 - \dfrac{x^4}{2!} + \dfrac{x^8}{4!} - \dfrac{x^{12}}{6!} + \cdots\right).$

6. $(x - 2)^2\Big\{c_1\Big[(x - 2) - \dfrac{(x - 2)^3}{3!} + \dfrac{(x - 2)^5}{5!} - \cdots\Big]$
$$+ c_2\Big[1 - \dfrac{(x - 2)^2}{2!} + \dfrac{(x - 2)^4}{4!} - \cdots\Big]\Big\}.$$

7. $\left[(x-1) + \dfrac{(x-1)^2}{1!} + \dfrac{(x-1)^3}{2!} + \dfrac{(x-1)^4}{3!} + \cdots\right](c_1 + c_2 \ln|x-1|).$

9. $c_1 x^2(1 + \frac{9}{5}x + \frac{49}{30}x^2 + \frac{209}{210}x^3 + \cdots)$

$$+ c_2 x^{-2}(3 + 3x + \tfrac{3}{2}x^2 + \tfrac{1}{2}x^3 - \tfrac{1}{5}x^5 + \cdots).$$

13. $c_1|x|^{-1/2}\left(x - \dfrac{x^3}{3!} + \dfrac{x^5}{5!} - \cdots\right) + c_2|x|^{-1/2}\left(1 - \dfrac{x^2}{2!} + \dfrac{x^4}{4!} - \cdots\right).$

14. $-x^3 + 9x^2 - 18x + 6.$

15. $c_1\left(x^3 - \dfrac{x^9}{3!} + \dfrac{x^{15}}{5!} - \cdots\right) + c_2\left(1 - \dfrac{x^6}{2!} + \dfrac{x^{12}}{4!} - \cdots\right).$

16. $c_1 y_1(x) + c_2\left[y_1(x)\ln|x| + \left(\dfrac{x}{2}\right)^2 - \dfrac{\frac{1}{1} + \frac{1}{2}}{(2!)^2}\left(\dfrac{x}{2}\right)^4 + \dfrac{\frac{1}{1} + \frac{1}{2} + \frac{1}{3}}{(3!)^2}\left(\dfrac{x}{2}\right)^6 - \cdots\right],$

$$y_1(x) = \sum_{0}^{\infty} (-1)^n \frac{1}{(n!)^2}\left(\frac{x}{2}\right)^{2n}.$$

4   The Bessel equation

The differential equation

(4.1)
$$x^2 y'' + x y' + (x^2 - n^2)y = 0 \qquad (n \geq 0)$$

is known as *Bessel's equation*. It is of considerable importance in mathematical physics. It is also of interest in its own right to mathematicians. We shall examine some of the simpler properties of solutions of this equation.

First, we note that Bessel's equation has a regular singular point at $x = 0$. The indicial equation associated with this point is

$$r^2 - n^2 = 0.$$

The indicial roots are $r = n,\ -n$. Thus, there exists a solution of (4.1) of the form $x^n P(x)$, where $P(x)$ is a convergent power series in powers of x, and $P(0) \neq 0$. If the substitution

(4.2)
$$y = x^n \sum_{0}^{\infty} a_m x^m$$

is made in equation (4.1), we find after the usual computation that

$$(2n + 1)a_1 = 0,$$

$$a_m = -\frac{a_{m-2}}{m(2n + m)} \qquad (m = 2, 3, \ldots).$$

Accordingly,

$$a_1 = a_3 = a_5 = \cdots = 0.$$

If a_0 is set equal to 1, we have

$$a_{2m} = \frac{(-1)^m}{2^{2m}m!(n+1)(n+2)\cdots(n+m)} \qquad (m = 1, 2, \ldots).$$

The solution (4.2) may then be written $(n \geq 0)$

$$(4.3) \quad y = x^n\left[1 - \frac{x^2}{2^2 1!(n+1)} + \frac{x^4}{2^4 2!(n+1)(n+2)} - \frac{x^6}{2^6 3!(n+1)(n+2)(n+3)} + \cdots\right].$$

This series has the test-ratio

$$(4.4) \qquad \left|\frac{x^2}{4(m+1)(n+m+1)}\right|,$$

which approaches zero as m becomes infinite. Thus the series in brackets in (4.3) converges for all values of x.

When n is an integer it is customary to multiply this solution by

$$c_n = \frac{1}{2^n n!}.$$

The resulting solution is called a Bessel function of order n and is designated by the symbol $J_n(x)$. It follows from the form of (4.1) that if n is not an integer, $J_{-n}(x)$ is also a solution of (4.1). Indeed, when n is not an integer, a second solution $J_{-n}(x)$ may be obtained simply by replacing n by $-n$ in (4.3). Further, in this case $J_n(x)$ and $J_{-n}(x)$ are linearly independent, for $J_{-n}(x) \not\equiv cJ_n(x)$ (c constant).

When n is an integer, the "second" solution may be obtained by the usual device of setting

$$y = J_n(x)v.$$

or by employing (3.14). The second solution will be found to involve $\ln |x|$, as we might expect from Theorem 3.1.

If the substitution

$$y = x^{-1/2}z$$

is made in (4.1), this equation becomes

$$(4.5) \qquad z'' + \left[1 - \frac{n^2 - \frac{1}{4}}{x^2}\right]z = 0.$$

When x is large this equation is approximately $z'' + z = 0$. This suggests that for x large the solutions $z(x)$ of (4.5) behave like solutions $\sin(x-a)$ of the differential equation

$$(4.6) \qquad v'' + v = 0.$$

That this is, in fact, the case requires analysis outside the limitations of this book.

When $n^2 = \frac{1}{4}$, equation (4.5) becomes

$$z'' + z = 0,$$

and [from (4.3), for example]

$$J_{1/2}(x) = c_1 \frac{\sin x}{\sqrt{x}},$$

$$J_{-1/2}(x) = c_2 \frac{\cos x}{\sqrt{x}}.$$

Note also that

$$J_0(x) = 1 - \frac{x^2}{2^2} + \frac{x^2}{2^2 4^2} - \frac{x^6}{2^2 4^2 6^2} + \cdots.$$

The general solution for $n = 0$ appears in the answer to Exercise 16 of Section 3. A graph of the functions $J_0(x)$ and $J_1(x)$ is given in Fig. 7.2.

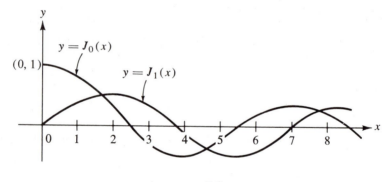

FIG. 7.2

Note that Fig. 7.2 suggests that the zeros of $J_1(x)$ separate those of $J_0(x)$, and vice versa. This is, in fact, the general situation, as the following theorem shows.

Theorem 4.1.   *Between every two consecutive positive zeros of $J_n(x)$ there is a zero of $J_{n-1}(x)$ and also of $J_{n+1}(x)$.*

The proof of the theorem is omitted.

Exercises

1.   It is customary to write

$$J_{1/2}(x) = \sqrt{\frac{2}{\pi}} \frac{\sin x}{\sqrt{x}},$$

$$J_{-1/2}(x) = \sqrt{\frac{2}{\pi}} \frac{\cos x}{\sqrt{x}}.$$

It can be shown that

$$J_{n+1}(x) = \frac{2n}{x} J_n(x) - J_{n-1}(x) \qquad (n \geq 0).$$

Use this formula to calculate $J_{3/2}(x)$, $J_{5/2}(x)$.

2.   Set $y = x^n w$ $(x > 0)$ in equation (4.1) and hence show that w is a solution of the differential equation

$$(x^{2n+1}w')' + x^{2n+1}w = 0.$$

Answers

1. $J_{3/2}(x) = \sqrt{\dfrac{2}{\pi}} \dfrac{1}{\sqrt{x}} \left(\dfrac{\sin x}{x} - \cos x \right);$

$J_{5/2}(x) = \sqrt{\dfrac{2}{\pi}} \dfrac{1}{\sqrt{x}} \left(\dfrac{3 \sin x}{x^2} - \dfrac{3 \cos x}{x} - \sin x \right).$

Volumes† have been written on Bessel functions. We shall limit ourselves, however, to stating a few more facts about them. All Bessel functions *oscillate*—that is, have an infinity of zeros—on the interval $(0, \infty)$. The distance between consecutive zeros tends to π as x becomes infinite [recall equation (4.5)]. When $n^2 > \frac{1}{4}$, the distance between consecutive zeros is always greater than π, while if $n^2 < \frac{1}{4}$, the distance between them is always less than π. Equation (4.5) shows that when $n^2 = \frac{1}{4}$, the distance between consecutive zeros is, of course, exactly π.

---

† A classic is G. N. Watson, *A Treatise on the Theory of Bessel Functions*, Cambridge University Press (1966).

8

Numerical methods

Both from a theoretical and a practical point of view the use of power series to approximate solutions of a differential equation is necessarily limited. In the first place, the differential equation must be analytic and possess an analytic solution. Even when this is true, power series

$$\sum a_n(x - a)^n$$

cannot be depended on to converge very rapidly except for values on its interval of convergence that are quite close to $x = a$. So other numerical methods have been devised. Such methods contribute little to a theory of the behavior of solutions of a differential equation, but in general they provide methods of approximating a specific solution with considerable accuracy.

1   The Cauchy polygon method

This is perhaps the simplest, and generally least accurate, method among those commonly employed to approximate a solution of a differential equation. An example will be sufficient to indicate the procedure.

Suppose that we wish to approximate the solution of the differential equation

$$\text{(1.1)} \qquad\qquad y' = -\frac{y}{1 + x}$$

that satisfies the condition

$$\text{(1.2)} \qquad\qquad y(0) = 1.$$

The fundamental existence theorem guarantees the existence of one, and only one, such solution, and that solution exists on the interval $(-1 < x < \infty)$.

Suppose, for simplicity, that we are interested in this solution only on the interval $[0, 1]$. The slope of this solution at the point $(0, 1)$ is readily computed from (1.1) to be -1. The interval $[0, 1]$ is subdivided into n subintervals. (We have broken the interval into five equal subintervals in Fig. 8.1. In

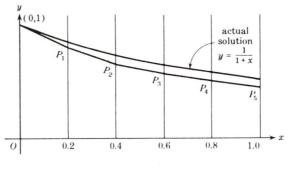

FIG. 8.1

general, the larger n is, the better is the approximation. Also, the subintervals need not be equal.) The straight-line segment through $(0, 1)$ with slope -1 has the equation

$$\text{(1.3)} \qquad\qquad y = 1 - x.$$

We take this straight line as the approximation to the solution on the interval $0 < x < 0.2$.

At $x = 0.2$, y is computed from (1.3) to be 0.8. The coordinates of P_1 in Fig. 8.1 are then $(0.2, 0.8)$. We now use equation (1.1) to compute the slope of the solution through P_1. We find this slope to be $-\frac{2}{3}$. The straight line through P_1 with slope $-\frac{2}{3}$ has the equation

(1.4) $$y = 0.8 - \tfrac{2}{3}(x - 0.2),$$

and this line is taken as the solution of the system (1.1) and (1.2) on the interval $0.2 < x < 0.4$. When $x = 0.4$, y is found from (1.4) to be 0.67. The coordinates of P_2 are then (0.4, 0.67). The process is continued, and points P_3, P_4, P_5 are determined consecutively. To two-place accuracy the coordinates of these points are found to be

$$P_3: (0.60, 0.57), \qquad P_4: (0.80, 0.50), \qquad P_5: (1.00, 0.44).$$

The precise solution of (1.1) and (1.2) is easily found to be

(1.5) $$y(x) = \frac{1}{1 + x}.$$

We give below a table of comparative values.

x	Actual y	Approx. y
0.00	1.00	1.00
0.20	0.83	0.80
0.40	0.71	0.67
0.60	0.62	0.57
0.80	0.56	0.50
1.00	0.50	0.44

It will be seen that the actual values of y are larger than the approximate values computed, and that the errors tend to accumulate. The former is true because the actual solution is concave upward on the interval. The latter observation is typical of the method.

It should be repeated that it is not necessary to proceed with equal subintervals to employ this method.

Exercises

Use the Cauchy polygon method to compute an approximate solution of each of the following differential systems on the interval indicated. Take n as given and use equal subintervals. Compute to two decimal places' accuracy. Draw the polygonal graph and the graph of the actual solution (use rather large-scale drawings).

1. $\dfrac{dy}{dx} = -\dfrac{y}{x+1}$, $y(0) = 1, 0 \le x \le 1; n = 10.$

2. (a) $\dfrac{dy}{dx} = \dfrac{1}{2y}$, $y(0) = 1, 0 \le x \le 1; n = 5.$

   (b) $\dfrac{dy}{dx} = \dfrac{1}{2y}$, $y(0) = 1, 0 \le x \le 1; n = 10.$

3. (a) $\dfrac{dy}{dx} = 1 + y$, $y(0) = 0, 0 \le x \le 1; n = 5.$

   (b) $\dfrac{dy}{dx} = 1 + y$, $y(0) = 0, 0 \le x \le 1; n = 10.$

4. (a) $y' = 2x - 2, y(0) = 1, 0 \le x \le 2, n = 4.$
   (b) $y' = 2x - 2, y(0) = 1, 0 \le x \le 2, n = 8.$

5. $y' = y^2 + 1, y(0) = 0, 0 \le x \le 1, n = 6.$

6. $y' = 1 - y^2, y(0) = 0, 0 \le x \le 1, n = 5.$

7. $y' = y^2 - 1, y(0) = 0, 0 \le x \le 1, n = 5.$

8. $y' = -\dfrac{x}{2y}$, $y(0) = \sqrt{2}, 0 \le x \le 2, n = 6.$

9. $y' = 1, y(0) = 0, 0 \le x \le 5, n = 4.$

2   The three-term Euler formula

This method is both reasonably simple and sufficiently accurate for many problems. Let us first illustrate its use on an example.

We wish to approximate the solution $y(x)$ of the system

(2.1) $$y' = 1 - x + 2y,$$

(2.2) $$y(0) = 1$$

on the interval $[0, 1]$. Equation (2.1) is a first-order linear differential equation, the solution of which, subject to condition (2.2), is readily calculated to be

(2.3) $$\tfrac{1}{2}x - \tfrac{1}{4} + \tfrac{5}{4}e^{2x}.$$

Knowing the exact answer will enable us to calculate the error in our use of approximate methods to which we now turn our attention.

Consider the first three terms of the Taylor expansion of the solution $y(x)$ about the point $x = 0$:

$$(2.4) \qquad y(h) = y(0) + h\frac{y'(0)}{1!} + h^2\frac{y''(0)}{2!} \qquad \text{(approximately)}.$$

Suppose we divide the interval $[0, 1]$ into $n = 5$ equal parts and let $h = 0.2$ be the length of the first subinterval. We wish to calculate the value $y_1 = y(h)$ from (2.4). From (2.2) we have $y(0) = 1$, and from (2.1) we have $y'(0) = 3$. Differentiating (2.1) we have

$$(2.5) \qquad\qquad y''(x) = -1 + 2y'(x);$$

accordingly, $y''(0) = 5$.

Equation (2.4) now yields

$$y_1 = 1 + 3h + 2.5h^2,$$

or, since $h = 0.2$, $y_1 = 1.700$.

The process is repeated with $y(x)$ expanded about the point $x = h$. Equation (2.4) becomes

$$(2.6) \qquad\qquad y(2h) = y(h) + h\frac{y'(h)}{1!} + h^2\frac{y''(h)}{2!}.$$

To compute $y_2 = y(2h)$ from this formula, we take

$$y_1 = y(h) = 1.700,$$
$$y'(h) = 1 - h + 2y_1 = 4.2 \qquad \text{[from (2.1)]},$$
$$y''(h) = -1 + 2y'(h) \qquad \text{[from (2.5)]}$$
$$= 7.4.$$

Equation (2.6) now yields

$$y_2 = 1.700 + 4.2h + 3.7h^2$$
$$= 2.688.$$

Similarly, we set

$$(2.7) \qquad\qquad y_3 = y_2 + h\frac{y'(2h)}{1!} + h^2\frac{y''(2h)}{2!},$$

and

$$y'(2h) = 1 - 2h + 2y_2 = 5.976,$$
$$y''(2h) = -1 + 2y'(2h) = 10.952.$$

From (2.7) we then have

$$y_3 = 2.688 + 5.976h + 5.476h^2$$
$$= 4.102.$$

Continuing the process we obtain

$$y_4 = 6.147,$$
$$y_5 = 9.126.$$

For comparison, here is a table of the actual and approximate values of the numbers y_m.

y_m	Actual	Approx.	Error E	Relative error E/y_m
y_1	1.715	1.700	0.015	0.009
y_2	2.732	2.688	0.044	0.016
y_3	4.200	4.102	0.098	0.023
y_4	6.341	6.147	0.194	0.031
y_5	9.486	9.126	0.360	0.038

The approximation is fairly good. Using a smaller value of h—and, hence, a larger value of n—would increase the accuracy. For example, for $n = 10$, $h = 0.1$, the approximate value of y_1 becomes 1.325, while the actual value of y_1 is 1.327, with an error of 0.002 and a relative error of 0.0015.

It is, of course, not necessary to require all steps to be of constant length h.

We are now prepared to put down formulas for the general case. Suppose our differential equation is

$$(2.8) \qquad y' = f(x, y),$$

where $f(x, y)$ and its partial derivatives $f_x(x, y)$ and $f_y(x, y)$ are continuous in that region of the xy-plane where we wish to approximate a solution $y(x)$ of (2.8) subject to the condition

$$y(x_0) = y_0.$$

From (2.8) we may calculate

$$y'' = f_x(x, y) + f_y(x, y)y'.$$

To pass from the point (x_{n-1}, y_{n-1}) to (x_n, y_n) we use the modified Taylor's formula

$$y_n = y_{n-1} + hy'_{n-1} + \frac{h^2}{2} y''_{n-1},$$

where $h = x_n - x_{n-1}$, and the numbers y'_{n-1} and y''_{n-1} are calculated, consecutively, from the equations

$$y'_{n-1} = f(x_{n-1}, y_{n-1}),$$
$$y''_{n-1} = f_x(x_{n-1}, y_{n-1}) + f_y(x_{n-1}, y_{n-1})y'_{n-1}.$$

Exercises

1. Do Exercise 1 of Section 1 by the method of this section.

2. Do Exercise 2 of Section 1 by the method of this section.

3. Do Exercise 3 of Section 1 by the method of this section.

4. Do Exercise 4 of Section 1 by the method of this section.

5. Do Exercise 5 of Section 1 by the method of this section.

6. Do Exercise 6 of Section 1 by the method of this section.

7. Do Exercise 7 of Section 1 by the method of the section.

8. Do Exercise 8 of Section 1 by the method of this section.

Use the method of this section to approximate the solutions of the following differential equations. Plot the points (x_i, y_i) and sketch the approximation curves.

9. $y' = -\dfrac{x^2 + y^2}{2xy}$, $y(1) = 3$, $1 \leq x \leq 3$; $n = 5$.

10. $y' = xy$, $y(1) = 1$, $1 \leq x \leq 2$; $n = 5$.

11. $y' = -2 + x + y$, $y(0) = 1$, $0 \leq x \leq 1$; $n = 5$.

12. $y' = 1 - 2x - 2y$, $y(-1) = 0$, $-1 \leq x \leq 1$; $n = 5$.

13. $y' = y - 2x$, $y(0) = 1$, $0 \leq x \leq 2$; $n = 5$.

14. $y' = 2x - 2$, $y(0) = 2$, $0 \leq x \leq 2$; $n = 4$.

3   The Runge-Kutta method

The most commonly used method for numerically approximating solutions of differential equations—particularly when a large computer is available†— and the most accurate, in general, is the so-called *Runge-Kutta method*. We shall limit ourselves to providing the basic formulas.‡

We consider the differential equation

$$y' = f(x, y),$$

and employing the notation of Section 2 we have

(3.1) $$y_{n+1} = y_n + \frac{h}{6}(k_1 + 2k_2 + 2k_3 + k_4),$$

where

(3.2)
$$k_1 = f(x_n, y_n),$$
$$k_2 = f(x_n + \tfrac{1}{2}h, y_n + \tfrac{1}{2}hk_1),$$
$$k_3 = f(x_n + \tfrac{1}{2}h, y_n + \tfrac{1}{2}hk_2),$$
$$k_4 = f(x_n + h, y_n + hk_3).$$

Formulas (3.1) and (3.2) provide a one-step process for passing from y_n to y_{n+1}. The numbers k_i in (3.2) will, of course, be different for each step— that is, they also depend on n. Note that the Runge-Kutta method does not involve the calculation of any partial derivatives. Like the methods of the previous sections, the Runge-Kutta method does not require the use of equal subintervals of common length h.

Let us apply this method to the example of Section 2.

We sought an approximation of the solution of the system

(3.3) $$y' = 1 - x + 2y,$$

(3.4) $$y(0) = 1.$$

We take $h = 0.2$, $n = 5$, and have $x_0 = 0$, $y_0 = 1$. From (3.2) we have then

$$k_1 = f(x_0, y_0) = 3,$$
$$k_2 = f(0.1, 1.3) = 3.5,$$
$$k_3 = f(0.1, 1.35) = 3.6,$$
$$k_4 = f(0.2, 1.72) = 4.24.$$

---

† Usually programs of Runge-Kutta computing are already available at large computing centers.

‡ The advent of the modern computer has generated great interest in numerical methods. Numerous volumes on the subject have been and are being written. See, for example, Peter Henrici, *Elements of Numerical Analysis*, John Wiley and Sons, New York (1964), 3rd printing (1967).

Then (3.1) yields

$$y_1 = 1 + \frac{0.2}{6}(3 + 7.0 + 7.2 + 4.24) = 1.715;$$

thus, $(x_1, y_1) = (0.2, 1.715)$.

We shall compute the next set of values of k_i and k_2 in full detail also. We have

$$k_1 = f(x_1, y_1) = f(0.2, 1.715) = 4.23,$$
$$k_2 = f(0.3, 2.138) = 4.976,$$
$$k_3 = f(0.3, 2.213) = 5.125,$$
$$k_4 = f(0.4, 2.740) = 5.080.$$

In this case, equation (3.1) yields

$$y_2 = 1.715 + \frac{0.2}{6}(4.23 + 9.952 + 10.250 + 5.080) = 2.732.$$

Continuing the process we obtain

$$y_3 = 4.200, \qquad y_4 = 6.341, \qquad y_5 = 9.485.$$

These results are very accurate indeed, as the following table shows.

y_n	*Actual*	*Method Section 2*	*Runge-Kutta*
y_1	1.715	1.700	1.715
y_2	2.732	2.688	2.732
y_3	4.200	4.102	4.200
y_4	6.341	6.147	6.341
y_5	9.486	9.126	9.485

Exercises

Do the exercises in Section 2 using the Runge-Kutta method. In all cases, take $n = 5$ and use equal subintervals.

4   The method of successive approximations; algebraic equations

The method of *successive approximations* is an exceedingly important practical method of solving algebraic equations as well as possibly the most important theoretical method for solving differential equations. Frequently, it is also useful as a practical method of solving differential equations.

Let us apply the method first to the trivial problem of solving the equation

(4.1) $$2x - 1 = x + 1.$$

The graphs of the functions $2x - 1$ and $x + 1$ appear in Fig. 8.2 and the steps in the solution of the problem may be followed from the diagram.

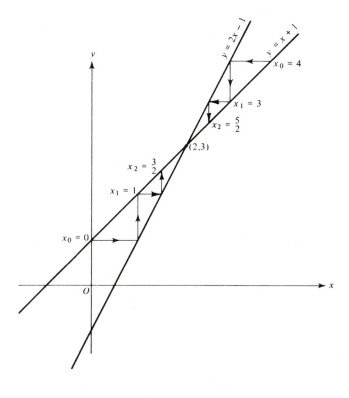

FIG. 8.2

First, we guess an approximation x_0 to the solution. Suppose our guess is $x_0 = 4$. We set this value in the right-hand member of (4.1) and compute a new approximation x_1 to x from (4.1), obtaining

$$2x_1 - 1 = 4 + 1,$$

or $x_1 = 3$. This value is put in the right-hand member of (4.1), and a new approximation $x_2 = \frac{5}{2}$ is obtained. It is clear from the geometry of the situation that the sequence

$$x_0, x_1, x_2, \ldots$$

will converge to the correct value, which is 2. This process is equivalent to computing numbers x_n from the *recursion* formulas

(4.2)
$$2x_n - 1 = x_{n-1} + 1,$$
$$x_0 = 4.$$

We note that if our original guess x_0 had been 5, 10, or 10^{10}, the process would have yielded eventually as good an approximation as needed to the true solution, the number 2. We observe from Fig. 8.2, however, that it is desirable for computational reasons to make as good a guess initially as possible.

Again, suppose that our initial guess had been $x_0 = 0$. The sequence of numbers x_n determined from (4.2) is

$$0, 1, \tfrac{3}{2}, \ldots.$$

From Fig. 8.2 we see that this sequence also will converge to the correct value 2.

If we had commenced by inserting the value 4 for x in the left-hand member of (4.1) and computed x_1 from the relation

$$2(4) - 1 = x_1 + 1,$$

we would have obtained $x_1 = 6$, and if we had continued in this fashion using, in effect, the recursion relation

(4.3)
$$2x_{n-1} - 1 = x_n + 1,$$

we would have obtained successively the values

$$4, 6, 10, 18, 34, \ldots.$$

It is easy to see that this sequence does not have a finite limit. It will also be apparent from Fig. 8.2 that the sequence so obtained does not approach the limit desired.

Although rules may be given which would enable the student to decide for a given equation which of the two recursion formulas, (4.2) or (4.3), to employ, in practice one more commonly either decides from a rough graph, or makes a guess. If the guess is wrong, this ordinarily becomes clear very quickly, and the correct procedure may then be followed.

If an error in computation is made, the result obtained amounts simply to a new guess, and the process is not invalidated. It may, indeed, actually be speeded up! If, for example, in computing x_1 in the first sequence we had made an error and obtained instead of 3 the value 2.5, we should have improved our approximating sequence.

Consider next the problem of approximating the solution of the equation

$$x = \cos x.$$

We try

(4.4)
$$x_n = \cos x_{n-1},$$
$$x_0 = .80$$

and obtain, successively, from a table of natural functions for angles given in radians (or from a hand calculator)

$$x_1 = .70, \quad x_2 = .76, \quad x_3 = .73, \quad x_4 = .745, \quad x_5 = .735, \ldots$$

It will be observed that the approximations $x_0, x_1, x_2, \ldots$ are alternately greater and less than the actual value of the solution. The average of any two approximations is better than either of them. A good approximation is then

$$x = .74.$$

The geometry involved is given in Fig. 8.3.

If we had set

$$\cos x_n = x_{n-1}, x_0 = .80$$

initially, we would have found

$$x_1 = .64, \quad x_2 = .88, \quad x_3 = .50, \ldots$$

and that the sequence was not converging. Note, however, that the average of the first two entries

$$\frac{.80 + .64}{2} = .72$$

provides in this instance a fairly good estimate of the answer.

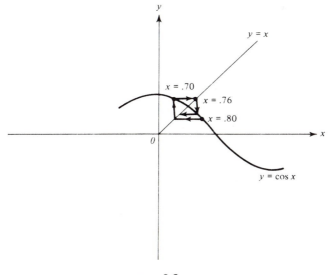

FIG. 8.3

Exercises

Solve the following equations by the method of successive approximations. Compute your answer to three decimal places.

1. $x - 3 = 2x + 2$.

2. $x = 1 + 1/x$ $(x > 0)$.

3. $x = 2 + 2/x$ $(x > 0)$.

4. $x^3 + 2x - 20 = 0$. (*Hint.* Rewrite in a convenient form.)

5. $x^x = 100$.

6. $2x - 1 = -2x + 1$. (Take $x_0 = 1$; examine the graphs, and make appropriate alterations of the process in the text.)

7. $x = 2 \sin x$ (root between 0 and π).

8. $\sec x = \cosh x$ (smallest positive root).

9.   Solve the following system for x (correct to two decimal places):

$$\sin \theta + x \cos \theta = 15,$$
$$\cos \theta + x \sin \theta = 10. \quad (0 < \theta < \pi/2).$$

[*Hint.* Write recursion formulas as

$$x_n = 15 \sec \theta_n - \tan \theta_n,$$
$$\csc \theta_{n+1} = \tfrac{1}{10}(x_n + \cot \theta_n),$$

and set $\theta_0 = \arctan(2/3)$ in the first of these equations.]

10.   Find the length x of the rectangle one foot wide that is inscribed in a rectangle 10 ft. × 20 ft. as indicated in the diagram.

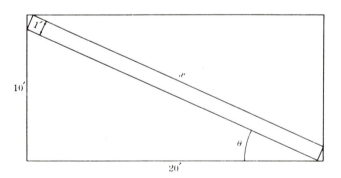

FIG. 8.4

Answers

4. 2.470.

5. 3.597.

7. 1.895.

8. 4.7300.

9. 17.12.

10. 21.58.

5 Differential Equations

We shall now apply the method to the differential system

$$y' - 2xy = 0,$$
$$y(0) = 1.$$

(5.1)

We rewrite (5.1) as

$$y = 1 + \int_0^x 2xy \, dx$$

from which we obtain the recursion formula

(5.2) $$y_n(x) = 1 + \int_0^x 2xy_{n-1}(x) \, dx.$$

Since $y(0) = 1$, a convenient guess for $y_0(x)$ is $y_0(x) = 1$. From (5.2) we obtain successively

$$y_1(x) = 1 + x^2,$$

$$y_2(x) = 1 + x^2 + \frac{x^4}{2},$$

$$y_3(x) = 1 + x^2 + \frac{x^4}{2} + \frac{x^6}{6},$$

$$\cdot \quad \cdot \quad \cdot \quad \cdot \quad \cdot \quad \cdot \quad \cdot \quad \cdot \quad \cdot$$

The sequence of functions so obtained can be shown to converge to the solution of (5.1) over every finite interval. They are clearly best suited to computation when $|x|$ is not too large.

The student will find it instructive to compute the sequence of functions $y_n(x)$ obtained from (5.2) for various selections of $y_0(x)$.

The method is also available for a system of differential equations. Consider, for example, the system

(5.3)

$$\frac{dy}{dx} = 3y - z,$$

$$\frac{dz}{dx} = 2y,$$

$$y(0) = 1, \qquad z(0) = 2.$$

As we shall learn in the next chapter, by a solution of such a system is meant a pair of functions

$$y(x)$$
$$z(x),$$

usually written in a column that, as a pair, satisfies the given differential equations in (5.3). To apply the method of successive approximations to (5.3) we rewrite the system in the form

$$y = 1 + \int_0^x (3y - z)\, dx,$$

$$z = 2 + \int_0^x 2y\, dx,$$

from which we obtain the recursion formulas

$$y_n(x) = 1 + \int_0^x [3y_{n-1}(x) - z_{n-1}(x)]\, dx,$$

$$z_n(x) = 2 + \int_0^x 2y_{n-1}(x)\, dx.$$

Natural choices of $y_0(x)$ and $z_0(x)$ are

$$y_0(x) = 1, \qquad z_0(x) = 2.$$

The completion of the problem is left to the student as an exercise.

As a practical method of solving differential equations, the use of the method of successive approximations is rather limited, owing to the difficulties which may arise in carrying out the indicated integrations. This is especially true when the differential equations are nonlinear.

Exercises

Solve the following differential systems by the method of successive approximations.

1. $y' - 2xy = 0$,

    $y(0) = 1$. Take $y_0(x)$ as $1 + x$ and determine $y_3(x)$.

2. $y' = 1 + y^2$,

    $y(0) = 0$. Take $y_0(x) = x$ and determine $y_2(x)$.

3. $\dfrac{dy}{dx} = 3y - z$,

    $\dfrac{dz}{dx} = 2y$,

    $y(0) = 1$, $z(0) = 2$. Take $y_0(x) = 1$, $z_0(x) = 2$. Find $y_3(x)$ and $z_3(x)$.

4. $\dfrac{dy}{dx} = z + x$,

    $\dfrac{dz}{dx} = y + x$,

    $y(0) = 2$, $z(0) = -2$. Take $y_0(x) = 2$, $z_0(x) = -2$. Find $y_3(x)$ and $z_3(x)$.

Answers

1. $y_3(x) = 1 + x^2 + \dfrac{x^4}{2} + \dfrac{x^6}{6} + \dfrac{8x^7}{105}$.

3. $y_3(x) = 1 + x + \dfrac{x^2}{2} + \dfrac{4x^3}{3}$,

    $z_3(x) = 2 + 2x + x^2 + \dfrac{4x^3}{3}$.

9

Systems of linear differential equations

1   The existence theorem

In this chapter we shall begin with the study of systems of first-order linear differential equations—that is, systems of the type

$$\frac{dx_1}{dt} = a_{11}(t)x_1 + a_{12}(t)x_2 + \cdots + a_{1n}(t)x_n,$$

$$\frac{dx_2}{dt} = a_{21}(t)x_1 + a_{22}(t)x_2 + \cdots + a_{2n}(t)x_n,$$

(1.1)

$$\cdot \quad \cdot \quad \cdot \quad \cdot \quad \cdot \quad \cdot \quad \cdot \quad \cdot \quad \cdot \quad \cdot \quad \cdot \quad ,$$

$$\frac{dx_n}{dt} = a_{n1}(t)x_1 + a_{n2}(t)x_2 + \cdots + a_{nn}(t)x_n.$$

We have taken t (which will often represent *time*) as the independent variable and the variables $x_1, x_2, \ldots, x_n$ as the unknowns. We shall suppose that the functions $a_{ij}(t)$ are continuous on an interval I.

By a *solution* $x(t)$ of such a system is meant a column of functions

(1.2)
$$
\begin{aligned}
&x_1(t) \\
&x_2(t) \\
&\quad\vdots \\
&x_n(t),
\end{aligned}
$$

each of class C' on I, that simultaneously satisfy (1.1).

The fundamental existence theorem, as it applies to (1.1), may be stated as follows.

Theorem 1.1. If $a_1, a_2, \ldots, a_n$ is an arbitrary set of constants and if t_0 is any point of the interval I, there exists one and only one solution of (1.1) with the property that

$$
x_1(t_0) = a_1, \qquad x_2(t_0) = a_2, \qquad \ldots, \qquad x_n(t_0) = a_n.
$$

This solution is defined on the entire interval I.

Corollary. A solution (1.2) of (1.1) with the property that at some point t_0 of I

$$
x_1(t_0) = x_2(t_0) = \cdots = x_n(t_0) = 0
$$

is identically zero on I; that is,

$$
x_1(t) \equiv x_2(t) \equiv \cdots \equiv x_n(t) \equiv 0.
$$

The proof of the corollary is easy and is left to the student. We note that if $u(t)$ and $v(t)$ are solutions, so is $u(t) + v(t)$, where by $u(t) + v(t)$ is meant the column

$$
\begin{aligned}
&u_1(t) + v_1(t) \\
&u_2(t) + v_2(t) \\
&\quad\vdots \\
&u_n(t) + v_n(t).
\end{aligned}
$$

So also is $cu(t)$, where by $cu(t)$ is meant the column

$$
\begin{aligned}
&cu_1(t) \\
&cu_2(t) \\
&\quad\vdots \\
&cu_n(t).
\end{aligned}
$$

It follows that if c and k are constants and $u(t)$ and $v(t)$ are solutions of (1.1), so also is the column $cu(t) + kv(t)$.

It is frequently desirable to transform a linear differential equation of order n into a system of the type (1.1). This can be accomplished as follows. Suppose the given differential equation is

$$(1.3) \qquad \frac{d^n x}{dt^n} = a_1(t)\frac{d^{n-1}x}{dt^{n-1}} + a_2(t)\frac{d^{n-2}x}{dt^{n-2}} + \cdots + a_{n-1}(t)\frac{dx}{dt} + a_n(t)x,$$

where the functions $a_i(t)$ are continuous on the interval I. If we set

$$(1.4) \qquad x_1 = x, \qquad x_2 = \frac{dx}{dt}, \qquad x_3 = \frac{d^2 x}{dt^2}, \qquad \ldots, \qquad x_n = \frac{d^{n-1}x}{dt^{n-1}},$$

equation (1.3) is transformed under (1.4) into the system

$$\frac{dx_1}{dt} = x_2,$$

$$\frac{dx_2}{dt} = x_3,$$

$$\cdots\cdots\cdots\cdots$$

$$\frac{dx_{n-1}}{dt} = x_n,$$

$$\frac{dx_n}{dt} = a_1(t)x_n + a_2(t)x_{n-1} + \cdots + a_n(t)x_1.$$

Thus, a system (1.1) may be regarded as a generalization of a single linear differential equation of order n.

For the remainder of this section we shall assume $n = 3$ in order to simplify slightly the presentation of the proofs. The alterations required for the general case will be quite evident.

We shall use dots to denote differentiation with respect to t; thus

$$\dot{x} = \frac{dx}{dt}, \qquad \ddot{x} = \frac{d^2 x}{dt^2}, \qquad \dot{a}(t) = \frac{d}{dt}\,a(t).$$

When $n = 3$, the system (1.1) becomes

$$\frac{dx_1}{dt} = a_{11}(t)x_1 + a_{12}(t)x_2 + a_{13}(t)x_3,$$

$$(1.1)' \qquad \frac{dx_2}{dt} = a_{21}(t)x_1 + a_{22}(t)x_2 + a_{23}(t)x_3,$$

$$\frac{dx_3}{dt} = a_{31}(t)x_1 + a_{32}(t)x_2 + a_{33}(t)x_3,$$

and a solution $x(t)$ becomes the column

$$x_1(t)$$
(1.2)'
$$x_2(t)$$
$$x_3(t).$$

Three solutions $u(t)$, $v(t)$, $w(t)$ are said to be *linearly dependent* if there exist constants c_1, c_2, c_3, not all zero, such that

(1.5) $$c_1 u(t) + c_2 v(t) + c_3 w(t) \equiv 0.$$

Condition (1.5) is to be interpreted as follows. Let the components of $u(t)$, $v(t)$, $w(t)$ be given, respectively, by the columns

$$u_1(t) \qquad v_1(t) \qquad w_1(t)$$
$$u_2(t) \qquad v_2(t) \qquad w_2(t)$$
$$u_3(t), \qquad v_3(t), \qquad w_3(t).$$

Then (1.5) is equivalent to the three conditions

$$c_1 u_1(t) + c_2 v_1(t) + c_3 w_1(t) \equiv 0,$$
(1.5)'
$$c_1 u_2(t) + c_2 v_2(t) + c_3 w_2(t) \equiv 0,$$
$$c_1 u_3(t) + c_2 v_3(t) + c_3 w_3(t) \equiv 0.$$

Solutions not linearly dependent are said to be *linearly independent*.

Theorem 1.2. There exist three linearly independent solutions of (1.1)'. *Every solution of the system can be written as a linear combination of these three solutions.*

The solutions $u(t)$, $v(t)$, $w(t)$ defined, respectively, by the conditions

$$u_1(t_0) = 1, \qquad v_1(t_0) = 0, \qquad w_1(t_0) = 0,$$
(1.6)
$$u_2(t_0) = 0, \qquad v_2(t_0) = 1, \qquad w_2(t_0) = 0,$$
$$u_3(t_0) = 0, \qquad v_3(t_0) = 0, \qquad w_3(t_0) = 1 \qquad (a \le t_0 \le b),$$

are readily shown to be linearly independent. For, if constants c_1, c_2, c_3, not all zero, exist such that (1.5)' holds, these conditions would hold in particular when $t = t_0$. But this is impossible. [Note that the solutions $u(t)$, $v(t)$, $w(t)$ defined by (1.6) *do* exist, because of Theorem 1.1.]

It remains to prove that every solution $x(t)$ of (1.1)' may be written as a linear combination of $u(t)$, $v(t)$, $w(t)$. We shall show that if $x_1(t)$, $x_2(t)$, $x_3(t)$ are the components of $x(t)$, then

$$x(t) \equiv x_1(t_0)u(t) + x_2(t_0)v(t) + x_3(t_0)w(t);$$

that is,

$$x_i(t) \equiv x_1(t_0)u_i(t) + x_2(t_0)v_i(t) + x_3(t_0)w_i(t) \qquad (i = 1, 2, 3).$$

First, note that $X(t)$, with components

$$X_i(t) = x_i(t) - [x_1(t_0)u_i(t) + x_2(t_0)v_i(t) + x_3(t_0)w_i(t)] \quad (i = 1, 2, 3),$$

is a solution of $(1.1)'$ for it is a linear combination of solutions. This solution has the property that

$$X_1(t_0) = X_2(t_0) = X_3(t_0) = 0.$$

Thus, by the corollary to Theorem 1.1, $X_i(t) \equiv 0$. The proof of the theorem is complete.

Exercises

1. Prove the corollary to Theorem 1.1.

2. Show that each of the columns

$$\begin{matrix} e^t & e^{2t} \\ e^t, & 3e^{2t} \end{matrix}$$

is a solution of the system

$$\frac{dx}{dt} = \frac{1}{2}x + \frac{1}{2}y,$$

$$\frac{dy}{dt} = -\frac{3}{2}x + \frac{5}{2}y.$$

Are these solutions linearly independent?

3. Given the system

$$\frac{dx_1}{dt} = \frac{t}{t-1}x_2 + \frac{1}{1-t}x_3,$$

$$\frac{dx_2}{dt} = \frac{1}{t(1-t)}x_2 + \frac{1}{t(t-1)}x_3,$$

$$\frac{dx_3}{dt} = \frac{2}{1-t}x_2 + \frac{2}{t-1}x_3,$$

show that the columns

$$\begin{matrix} 1 & t & 1 \\ 0 & 1 & t \\ 0, & 1, & t^2 \end{matrix}$$

are linearly independent solutions. Hence, show that the columns

$$
\begin{array}{cccc}
t + 2 & 0 & t - 1 & t + 1 \\
t + 1 & t & 1 - t & 1 \\
t^2 + 1, & t^2, & (1 - t)(1 + t), & 1
\end{array}
$$

are also solutions. Are they linearly independent?

4.   Show that the four solutions

$$
\begin{array}{cccc}
u\colon & v\colon & w\colon & z\colon \\
1 & t & 1 & t + 2 \\
0 & 1 & t & t + 1 \\
0, & 1, & t^2, & t^2 + 1
\end{array}
$$

of the system in Exercise 3 are linearly dependent by finding four constants c_1, c_2, c_3, c_4, not all zero, such that

$$
c_1 u + c_2 v + c_3 w + c_4 z \equiv 0.
$$

2   Fundamental systems of solutions

In this, as in the preceding section, we shall devote our attention to the system $(1.1)'$ as a representative of the more general system (1.1).

Let the columns

$$
\begin{array}{ccc}
u_1(t) & v_1(t) & w_1(t) \\
u_2(t) & v_2(t) & w_2(t) \\
u_3(t), & v_3(t), & w_3(t)
\end{array}
$$

be solutions $u(t)$, $v(t)$, $w(t)$, respectively, of $(1.1)'$ and form the determinant

$$
\Delta(t) = \begin{vmatrix} u_1(t) & v_1(t) & w_1(t) \\ u_2(t) & v_2(t) & w_2(t) \\ u_3(t) & v_3(t) & w_3(t) \end{vmatrix}.
$$

Note that its derivative

$$
\dot{\Delta}(t) = \begin{vmatrix} \dot{u}_1(t) & \dot{v}_1(t) & \dot{w}_1(t) \\ u_2(t) & v_2(t) & w_2(t) \\ u_3(t) & v_3(t) & w_3(t) \end{vmatrix} + \begin{vmatrix} u_1(t) & v_1(t) & w_1(t) \\ \dot{u}_2(t) & \dot{v}_2(t) & \dot{w}_2(t) \\ u_3(t) & v_3(t) & w_3(t) \end{vmatrix}
$$

(2.1)

$$
+ \begin{vmatrix} u_1(t) & v_1(t) & w_1(t) \\ u_2(t) & v_2(t) & w_2(t) \\ \dot{u}_3(t) & \dot{v}_3(t) & \dot{w}_3(t) \end{vmatrix}.
$$

From $(1.1)'$ we see that

$$\dot{u}_i = a_{i1}u_1 + a_{i2}u_2 + a_{i3}u_3,$$

(2.2) $$\dot{v}_i = a_{i1}v_1 + a_{i2}v_2 + a_{i3}v_3,$$

$$\dot{w}_i = a_{i1}w_1 + a_{i2}w_2 + a_{i3}w_3 \qquad (i = 1, 2, 3).$$

If the dot quantities in the determinants (2.1) are replaced by the equivalent quantities given by (2.2), we shall see that

(2.3) $$\dot{\Delta}(t) \equiv [a_{11}(t) + a_{22}(t) + a_{33}(t)]\Delta(t).$$

To verify this, note that the first determinant in (2.1) becomes

$$\Delta_1 = \begin{vmatrix} a_{11}u_1 + a_{12}u_2 + a_{13}u_3 & a_{11}v_1 + a_{12}v_2 + a_{13}v_3 & a_{11}w_1 + a_{12}w_2 + a_{13}w_3 \\ u_2 & v_2 & w_2 \\ u_3 & v_3 & w_3 \end{vmatrix}.$$

If a_{12} times the second row and a_{13} times the third row of this determinant are subtracted term-by-term from the first row, the value of the determinant is unchanged and

$$\Delta_1 = \begin{vmatrix} a_{11}u_1 & a_{11}v_1 & a_{11}w_1 \\ u_2 & v_2 & w_2 \\ u_3 & v_3 & w_3 \end{vmatrix} = a_{11}\Delta(t).$$

In a similar fashion, the second and third determinants in (2.1) can be shown to have the values $a_{22}\Delta(t)$ and $a_{33}\Delta(t)$, respectively. The result (2.3) then follows.

We note from (2.3) that $\Delta(t)$ is a solution of the first-order linear differential equation

$$\dot{\Delta} = [a_{11}(t) + a_{22}(t) + a_{33}(t)]\Delta.$$

It follows that $\Delta(t)$ is either never zero or is identically zero. Indeed,

$$\Delta(t) = \Delta(t_0)e^{\int_{t_0}^{t} [a_{11}(t) + a_{22}(t) + a_{33}(t)] \, dt},$$

where t_0 is any point of the interval I.

A system of three solutions of $(1.1)'$ with the property that $\Delta(t) \not\equiv 0$ is called a *fundamental system* of solutions.

Theorem 2.1. *A necessary and sufficient condition that three solutions of* $(1.1)'$ *be linearly independent is that they form a fundamental system.*

To prove this theorem we require an extension of Lemma 6.1 of Chapter 3. It is the following theorem from linear algebra.

Lemma 2.1. *A necessary and sufficient condition that there exist constants* $c_1, c_2, \ldots, c_n$, *not all zero, such that*

$$a_{11}c_1 + a_{12}c_2 + \cdots + a_{1n}c_n = 0,$$
$$a_{21}c_1 + a_{22}c_2 + \cdots + a_{2n}c_n = 0,$$
$$\cdots\cdots\cdots\cdots\cdots\cdots\cdots\cdots\cdots\cdots\cdots\cdots$$
$$a_{n1}c_1 + a_{n2}c_2 + \cdots + a_{nn}c_n = 0$$

is that the determinant

$$\begin{vmatrix} a_{11} & a_{12} & \cdots & a_{1n} \\ a_{21} & a_{22} & \cdots & a_{2n} \\ \cdots\cdots\cdots\cdots\cdots\cdots\cdots \\ a_{n1} & a_{n2} & \cdots & a_{nn} \end{vmatrix} = 0.$$

A proof of Theorem 2.1 is now readily made. Let $u(t)$, $v(t)$, $w(t)$, with components

$$(2.4) \qquad \begin{matrix} u_1(t), & v_1(t), & w_1(t) \\ u_2(t) & v_2(t) & w_2(t) \\ u_3(t), & v_3(t), & w_3(t) \end{matrix}$$

be three solutions of the system $(1.1)'$. If constants c_1, c_2, c_3, not all zero, exist such that

$$(2.5) \qquad c_1 u(t) + c_2 v(t) + c_3 w(t) \equiv 0 \qquad \text{(on } I),$$

then, in particular, at any arbitrary point $t = t_0$ of I

$$c_1 u(t_0) + c_2 v(t_0) + c_3 w(t_0) = 0;$$

that is, constants c_1, c_2, c_3, not all zero, exist such that

$$(2.6) \qquad \begin{aligned} c_1 u_1(t_0) + c_2 v_1(t_0) + c_3 w_1(t_0) &= 0, \\ c_1 u_2(t_0) + c_2 v_2(t_0) + c_3 w_2(t_0) &= 0, \\ c_1 u_3(t_0) + c_2 v_3(t_0) + c_3 w_3(t_0) &= 0. \end{aligned}$$

It follows from Lemma 2.1 that the determinant

$$(2.7) \qquad \Delta(t_0) = \begin{vmatrix} u_1(t_0) & v_1(t_0) & w_1(t_0) \\ u_2(t_0) & v_2(t_0) & w_2(t_0) \\ u_3(t_0) & v_3(t_0) & w_3(t_0) \end{vmatrix} = 0.$$

But $t = t_0$ was an arbitrary point of I. It follows that $\Delta(t) \equiv 0$.

Conversely, suppose that $\Delta(t) \equiv 0$. Then, if $t = t_0$ is any point of I, $\Delta(t_0) = 0$. By Lemma 2.1, there then exist constants c_1, c_2, c_3, not all zero, such that equations (2.6) hold. Consider the solution

$$z(t) = c_1 u(t) + c_2 v(t) + c_3 w(t)$$

of $(1.1)'$ with this choice of constants. Then

$$z(t_0) = 0,$$

and it follows from the corollary to Theorem 1.1 that $z(t) \equiv 0$.

The proof of the theorem is complete.

Theorem 2.2. If $u(t)$, $v(t)$, $w(t)$ are any three linearly independent solutions of $(1.1)'$, every solution of the system may be written as a linear combination of these three solutions.

Corollary. Any four solutions of $(1.1)'$ are linearly dependent.

The proof of this theorem is left to the student.

Exercises

1. Compute $\Delta(t)$ for the system of solutions given for Exercise 2, Section 1, and thus verify that

    $$\dot{\Delta}(t) \equiv [a_{11}(t) + a_{22}(t)]\Delta(t).$$

2. Compute $\Delta(t)$ for one of the sets of three solutions given for Exercise 3, Section 1, and thus verify that

    $$\dot{\Delta}(t) \equiv [a_{11}(t) + a_{22}(t) + a_{33}(t)]\Delta(t).$$

    Does this example invalidate the conclusion of the text that $\Delta(t)$ is either identically zero or never zero ?

3.* Prove Theorem 2.2.

3   Linear systems with constant coefficients

When the coefficients $a_{ij}(t)$ in equations (1.1) are constants, we may write these equations as

$$
\begin{aligned}
\dot{x}_1 &= a_{11}x_1 + a_{12}x_2 + \cdots + a_{1n}x_n, \\
\dot{x}_2 &= a_{21}x_1 + a_{22}x_2 + \cdots + a_{2n}x_n, \\
&\cdots\cdots\cdots\cdots\cdots\cdots\cdots\cdots\cdots\cdots \\
\dot{x}_n &= a_{n1}x_1 + a_{n2}x_2 + \cdots + a_{nn}x_n.
\end{aligned}
$$

(3.1)

Recall that by a solution of such a system is meant a column of n functions

$$
\begin{matrix}
x_1(t) \\
x_2(t) \\
\vdots \\
x_n(t)
\end{matrix}
$$

that satisfies these equations simultaneously. To solve such a system we attempt to find a solution of the form

$$
\begin{matrix}
A_1 e^{\lambda t} \\
A_2 e^{\lambda t} \\
\vdots \\
A_n e^{\lambda t},
\end{matrix}
$$

(3.2)

where $\lambda, A_1, A_2, \ldots, A_n$ are constants.

Substituting (3.2) in (3.1) leads to the equations

$$
\begin{aligned}
(a_{11} - \lambda)A_1 + a_{12}A_2 + \cdots + a_{1n}A_n &= 0, \\
a_{21}A_1 + (a_{22} - \lambda)A_2 + \cdots + a_{2n}A_n &= 0, \\
&\cdots\cdots\cdots\cdots\cdots\cdots\cdots\cdots\cdots\cdots \\
a_{n1}A_1 + a_{n2}A_2 + \cdots + (a_{nn} - \lambda)A_n &= 0
\end{aligned}
$$

(3.3)

for the determination of $A_1, \ldots, A_n$. We seek a solution $A_1, A_2, \ldots, A_n$ of (3.3) where the A_i are not all zero. According to Lemma 2.1 of the preceding section, a necessary and sufficient condition for such a solution to exist is that the determinant

(3.4)
$$
\begin{vmatrix}
(a_{11} - \lambda) & a_{12} & \cdots & a_{1n} \\
a_{21} & (a_{22} - \lambda) & \cdots & a_{2n} \\
\cdots\cdots & \cdots\cdots & & \cdots\cdots \\
a_{n1} & a_{n2} & \cdots & (a_{nn} - \lambda)
\end{vmatrix} = 0.
$$

Equation (3.4) is a polynomial equation in λ of degree n and has, then, n roots. It is known as the *characteristic equation* of the system (3.1).

When λ_1 is a root of this equation, a solution $A_1, A_2, \ldots, A_n$, not all zero, of (3.3) exists and the column

$$
\begin{aligned}
&A_1 e^{\lambda_1 t} \\
&A_2 e^{\lambda_1 t} \\
&\vdots \\
&A_n e^{\lambda_1 t}
\end{aligned}
$$

will be of a solution of (3.1).

As usual, the n roots λ of the characteristic equation may be all real and distinct, or some or all roots may be complex numbers, or some of either of these types may be multiple roots.

We shall consider first the case $n = 2$. Then, the system (3.1) becomes

(3.5)
$$
\dot{x}_1 = a_{11}x_1 + a_{12}x_2,
$$
$$
\dot{x}_2 = a_{21}x_1 + a_{22}x_2,
$$

while (3.3) and (3.4) become, respectively,

(3.6)
$$
(a_{11} - \lambda)A_1 + a_{12}A_2 = 0,
$$
$$
a_{21}A_1 + (a_{22} - \lambda)A_2 = 0,
$$

and

(3.7)
$$
\begin{vmatrix} a_{11} - \lambda & a_{12} \\ a_{21} & a_{22} - \lambda \end{vmatrix} = 0.
$$

In this case the characteristic equation (3.7) is the quadratic equation

(3.8)
$$
\lambda^2 - (a_{11} + a_{22})\lambda + (a_{11}a_{22} - a_{12}a_{21}) = 0.
$$

The two roots of equation (3.8) are either real and unequal, or real and equal, or conjugate imaginaries. In the following example, the characteristic roots are real and unequal.

Example.   Solve the system

(3.9)
$$
\frac{dx_1}{dt} = \frac{1}{2}x_1 + \frac{1}{2}x_2,
$$
$$
\frac{dx_2}{dt} = -\frac{3}{2}x_1 + \frac{5}{2}x_2.
$$

Let us attempt to determine constants A, B, and λ such that the column

$$Ae^{\lambda t}$$
$$Be^{\lambda t}$$

is a solution of (3.9). Equations (3.6) become

(3.10)

$$(\tfrac{1}{2} - \lambda)A + \tfrac{1}{2}B = 0,$$

$$-\tfrac{3}{2}A + (\tfrac{5}{2} - \lambda)B = 0,$$

while the characteristic equation (3.7) becomes

(3.11)

$$\begin{vmatrix} \tfrac{1}{2} - \lambda & \tfrac{1}{2} \\ -\tfrac{3}{2} & \tfrac{5}{2} - \lambda \end{vmatrix} = 0,$$

or

$$\lambda^2 - 3\lambda + 2 = 0,$$

the roots of which are 1 and 2. If we set $\lambda = 1$ in (3.10), we obtain

(3.12)

$$-\frac{A}{2} + \frac{B}{2} = 0,$$

$$-\frac{3A}{2} + \frac{3B}{2} = 0.$$

These simultaneous equations are consistent, and we may choose as solutions of them any pair of numbers A and B, not both zero, satisfying them. An obvious choice is $A = B = 1$. This leads to the solution

$$e^t$$
$$e^t.$$

If we had chosen any other pair of numbers A, B satisfying (3.12), we would simply have obtained a constant times this solution.

In a similar way the root $\lambda = 2$ leads to the equations

$$-\frac{3A}{2} + \frac{B}{2} = 0,$$

$$-\frac{3A}{2} + \frac{B}{2} = 0.$$

A nontrivial solution of this system is $A = 1$, $B = 3$, and the corresponding solution of (3.9) is

$$e^{2t}$$
$$3e^{2t}.$$

It is easy to verify that these two solutions are linearly independent, and accordingly, every solution of (3.9) may be written in the form

(3.13)
$$c_1 e^t + c_2 e^{2t}$$
$$c_1 e^t + 3c_2 e^{2t},$$

where c_1 and c_2 are constants. Since, conversely, for every choice of the constants c_1 and c_2 the column (3.13) is a solution, it is appropriate to refer to (3.13) as the *general solution* of (3.9).

In the next example, the characteristic roots are real and equal.

Example. Solve the system

(3.14)
$$\frac{dx_1}{dt} = 4x_2,$$
$$\frac{dx_2}{dt} = -9x_1 + 12x_2.$$

The substitution in (3.14) of the column

$$A e^{\lambda t}$$
$$B e^{\lambda t}$$

leads in this case to the equations

(3.15)
$$-\lambda A + 4B = 0,$$
$$-9A + (12 - \lambda)B = 0,$$

and to the characteristic equation

(3.16)
$$\begin{vmatrix} -\lambda & 4 \\ -9 & 12 - \lambda \end{vmatrix} = 0.$$

The roots of (3.16) are seen to be $\lambda = 6, 6$. Setting $\lambda = 6$ in (3.15), we determine the nontrivial solution $A = 2$, $B = 3$, and the corresponding solution

(3.17)
$$2e^{6t}$$
$$3e^{6t}$$

of the system (3.14). Since the roots of (3.16) are equal, we cannot employ the second root of this quadratic equation to obtain a second linearly independent solution of (3.14). The situation here is similar to that which arose when we solved a linear differential equation of second order whose corresponding indicial equation possessed a repeated root (see Chapter 3, Section 2). This similarity might suggest that a second solution of (3.14) could be obtained by multiplying the solution (3.17) by t:

$$Cte^{2t}$$

$$Dte^{2t}.$$

It is easy to see, however, by testing these functions in (3.14), that they do not provide a solution. It is true, however, that a second solution of the form

(3.18)
$$(2t + \alpha)e^{6t}$$
$$(3t + \beta)e^{6t} \qquad (\alpha, \beta \text{ constants})$$

always exists.

Substitution of (3.18) in (3.14) leads to the identities

$$3\alpha - 2\beta = -1,$$
$$3\alpha - 2\beta = -1,$$

and we may choose $\alpha = \beta = -1$ to obtain a second solution

(3.19)
$$(2t - 1)e^{6t}$$
$$(3t - 1)e^{6t}.$$

The determinant

$$\begin{vmatrix} 2e^{6t} & (2t - 1)e^{6t} \\ 3e^{6t} & (3t - 1)e^{6t} \end{vmatrix} = e^{12t} \neq 0;$$

consequently, the solutions (3.17) and (3.19) are linearly independent.

The choice of the factors $(2t + \alpha)$ and $(3t + \beta)$ in (3.18) is a consequence of the fact that if λ_1 is a double root of the characteristic equation, and if one solution of the pair of differential equations is

$$he^{\lambda_1 t}$$
$$ke^{\lambda_1 t},$$

there is a second solution of the form

(3.20)
$$(ht + \alpha)e^{\lambda_1 t}$$
$$(kt + \beta)e^{\lambda_1 t}.$$

This observation extends to systems of n equations ($n = 3, 4, \ldots$). A proof of this principle is not difficult, but it is relatively long and is omitted. Of course, when the substitution of (3.20) in the given pair of differential equations permits the determination of α and β, one has clearly determined a second linearly independent solution.

Further, when λ_1 is a triple root, for example, of a system of n equations and one solution is

$$A_1 e^{\lambda_1 t}$$
$$A_2 e^{\lambda_1 t}$$
$$\vdots$$
$$A_n e^{\lambda_1 t},$$

there are also linearly independent solutions of the form

(3.21)
$$
\begin{array}{ll}
(A_1 t + \alpha_1) e^{\lambda_1 t} & (A_1 t^2 + \beta_1 t + \gamma_1) e^{\lambda_1 t} \\
(A_2 t + \alpha_2) e^{\lambda_1 t} & (A_2 t^2 + \beta_2 t + \gamma_2) e^{\lambda_1 t} \\
\quad\vdots & \quad\vdots \\
(A_n t + \alpha_n) e^{\lambda_1 t} & (A_n t^2 + \beta_n t + \gamma_n) e^{\lambda_1 t},
\end{array}
$$

where the α_i, β_i, γ_i ($i = 1, 2, \ldots, n$) are constants to be determined. And so on.

In the next example, the characteristic roots are conjugate imaginary numbers.

Example. Solve the system

(3.22)
$$\frac{dx_1}{dt} = 3x_1 + 2x_2,$$

$$\frac{dx_2}{dt} = -x_1 + x_2.$$

The substitution in (3.22) of the column

$$A e^{\lambda t}$$
$$B e^{\lambda t}$$

leads to the equations

(3.23)
$$(3 - \lambda)A + 2B = 0,$$
$$-A + (1 - \lambda)B = 0,$$

and the characteristic equation

(3.24)
$$
\begin{vmatrix} 3 - \lambda & 2 \\ -1 & 1 - \lambda \end{vmatrix} = 0.
$$

The roots λ of (3.24) are $2 + i$ and $2 - i$. The substitution of $\lambda = 2 + i$ in (3.23) leads to the nontrivial solution $A = 1 + i$, $B = -1$ of (3.23) and to the solution

(3.25)
$$(1 + i)e^{(2 + i)t}$$
$$-e^{(2 + i)t}$$

of (3.22). The solution (3.25) can be rewritten, using the familiar relationship

$$e^{(a+ib)t} = e^{at}(\cos bt + i \sin bt),$$

in the form

$$e^{2t}[(\cos t - \sin t) + i(\cos t + \sin t)]$$
$$-e^{2t}[\cos t + i \sin t].$$

The real part of this solution,

(3.26)
$$e^{2t}(\cos t - \sin t)$$
$$-e^{2t}\cos t,$$

must then also be a solution. So also is the imaginary part

(3.27)
$$e^{2t}(\cos t + \sin t)$$
$$-e^{2t}\sin t.$$

The solutions (3.26) and (3.27) may, of course, be tested directly in (3.22). The computation of

$$\Delta(t) = \begin{vmatrix} e^{2t}(\cos t - \sin t) & e^{2t}(\cos t + \sin t) \\ -e^{2t}\cos t & -e^{2t}\sin t \end{vmatrix} = e^{4t}$$

shows that the solutions (3.26) and (3.27) are linearly independent.

The general solution of (3.22) is, accordingly,

$$c_1 e^{2t}(\cos t - \sin t) + c_2 e^{2t}(\cos t + \sin t)$$
$$-c_1 e^{2t}\cos t - c_2 e^{2t}\sin t.$$

Since the root $\lambda = 2 + i$ of (3.24) yielded the general solution of (3.22), we can gain nothing by the substitution of the second root $\lambda = 2 - i$ of (3.24). The substitution $\lambda = 2 - i$ leads, in fact, to the same solution as the one we have just found.

To repeat then, the three examples above illustrate all the possibilities for a system (3.5), since the roots λ of the corresponding determinantal equation

$$\begin{vmatrix} a_{11} - \lambda & a_{12} \\ a_{21} & a_{22} - \lambda \end{vmatrix} = 0$$

fall into one of the three following categories:

1. roots real and distinct;
2. roots real and equal;
3. roots conjugate imaginary.

The substitution of

$$Ae^{\lambda t}$$
$$Be^{\lambda t}$$
$$Ce^{\lambda t}$$

in a system of three linear differential equations with constant coefficients leads to similar analysis.

Example.   Find the solution of the system

$$\dot{x}_1 = 3x_1 + 2x_2,$$
$$\dot{x}_2 = -x_1 + x_2,$$
$$x_1(0) = 0, \qquad x_2(0) = 1.$$

This is the system (3.22) plus boundary conditions. According to the fundamental existence theorem, these conditions will determine one solution uniquely. We have found that every solution $x(t)$ of the given pair of differential equations may be written in the form

$$x_1(t) = c_1 e^{2t}(\cos t - \sin t) + c_2 e^{2t}(\cos t + \sin t)$$
$$x_2(t) = -c_1 e^{2t}\cos t - c_2 e^{2t}\sin t.$$

From the boundary conditions we have

$$0 = c_1 + c_2,$$
$$1 = -c_1;$$

accordingly, $c_1 = -1$, $c_2 = 1$, and the solution we seek is the column

$$-e^{2t}(\cos t - \sin t) + e^{2t}(\cos t + \sin t)$$
$$e^{2t}\cos t - e^{2t}\sin t.$$

Inhomogeneous equations.   Consider the *inhomogeneous* system

(3.28)
$$\dot{x}_1 = \tfrac{1}{2}x_1 + \tfrac{1}{2}x_2,$$
$$\dot{x}_2 = -\tfrac{3}{2}x_1 + \tfrac{5}{2}x_2 + 20 \cos t.$$

The *corresponding homogeneous* system [see equations (3.9)]

$$\dot{x}_1 = \tfrac{1}{2}x_1 + \tfrac{1}{2}x_2,$$
$$\dot{x}_2 = -\tfrac{3}{2}x_1 + \tfrac{5}{2}x_2$$

has the general solution

(3.29)
$$x_1 = c_1 e^t + c_2 e^{2t},$$
$$x_2 = c_1 e^t + 3c_2 e^{2t}.$$

We seek a particular solution of the system (3.28). To find it, we try to determine constants a, b, h, and k so that the column

(3.30)
$$x_1 = a \sin t + b \cos t,$$
$$x_2 = h \sin t + k \cos t$$

is a solution. Substituting (3.30) in (3.28) we find that

$$a = -3, \qquad b = 1, \qquad h = 1, \qquad k = -7,$$

and a particular solution of (3.28) is then

(3.31)
$$x_1 = -3 \sin t + \cos t,$$
$$x_2 = \sin t - 7 \cos t.$$

The treatment of the inhomogeneous single linear differential equation (Chapter 3, Section 3) suggests that the general solution of (3.28) ought to be the "sum" of (3.31) and (3.29); that is,

$$x_1 = -3 \sin t + \cos t + c_1 e^t + c_2 e^{2t},$$
$$x_2 = \sin t - 7 \cos t + c_1 e^t + 3c_2 e^{2t}.$$

This is indeed the case, here and in general. We shall, however, omit the proof.

Exercises

Find the general solutions of the given systems of differential equations.

1. $\dfrac{dx_1}{dt} = x_2, \qquad \dfrac{dx_2}{dt} = -2x_1 + 3x_2.$

2. $\dfrac{dx_1}{dt} = -3x_1 + 4x_2, \qquad \dfrac{dx_2}{dt} = -2x_1 + 3x_2.$

3. $\dfrac{dx_1}{dt} = x_2, \qquad \dfrac{dx_2}{dt} = -x_1 + 2x_2.$

4. $\dfrac{dx_1}{dt} = 4x_1 - x_2, \qquad \dfrac{dx_2}{dt} = 4x_1.$

5. $\dfrac{dx_1}{dt} = 12x_1 - 5x_2, \qquad \dfrac{dx_2}{dt} = 30x_1 - 13x_2.$

6. $\dfrac{dx_1}{dt} = -2x_2, \qquad \dfrac{dx_2}{dt} = x_1 + 2x_2.$

7. $\dfrac{dx_1}{dt} = x_1 - 2x_2, \qquad \dfrac{dx_2}{dt} = 4x_1 + 5x_2.$

8. $\dfrac{dx}{dt} = 4x - 2z, \qquad \dfrac{dz}{dt} = 4x.$

Find a fundamental system of solutions for each of the following systems.

9. $\dfrac{dx_1}{dt} = 2x_1 + x_2 - 2x_3,$

$\dfrac{dx_2}{dt} = 3x_2 - 2x_3,$

$\dfrac{dx_3}{dt} = 3x_1 + x_2 - 3x_3.$

10. $\dfrac{dx_1}{dt} = x_2,$

$2\dfrac{dx_2}{dt} = -3x_1 + 6x_2 - x_3,$

$\dfrac{dx_3}{dt} = -x_1 + x_2 + x_3.$

11. $\dfrac{dx_1}{dt} = 3x_1 - 3x_2 + x_3,$

$\dfrac{dx_2}{dt} = 2x_1 - x_2,$

$\dfrac{dx_3}{dt} = x_1 - x_2 + x_3.$

12. $\dfrac{dx_1}{dt} = x_2,$

$\dfrac{dx_2}{dt} = x_3,$

$\dfrac{dx_3}{dt} = 4x_1 - 3x_3.$

13. $\dfrac{dx_1}{dt} = x_2,$

$\dfrac{dx_2}{dt} = x_3,$

$\dfrac{dx_3}{dt} = 8x_1 - 12x_2 + 6x_3.$

14. $\dot{x}_1 = x_2,$
$\dot{x}_2 = x_3,$
$\dot{x}_3 = -x_1 + x_2 + x_3.$

Also, find the solution that satisfies the conditions
$$x_1(0) = 0, \qquad x_2(0) = 1, \qquad x_3(0) = 0.$$

15. $\dot{x}_1 = 7x_2 - 6x_3,$
$\dot{x}_2 = -x_1 + 4x_2,$
$\dot{x}_3 = x_2 - 2x_3.$

Also, find the solution that satisfies the conditions
$$x_1(0) = 0, \qquad x_2(0) = 0, \qquad x_3(0) = 1.$$

16. $\dot{x}_1 = x_1 + x_2,$
$\dot{x}_2 = x_2,$
$\dot{x}_3 = x_3.$

Also, find the solution that satisfies the conditions
$$x_1(0) = 1, \qquad x_2(0) = 1, \qquad x_3(0) = 1.$$

17. $\dot{x}_1 = 9x_1 - 6x_2 + 2x_3,$
$\dot{x}_2 = -6x_1 + 8x_2 - 4x_3,$
$\dot{x}_3 = 2x_1 - 4x_2 + 4x_3.$

Also, find the solution that satisfies the conditions
$$x_1(0) = 0, \qquad x_2(0) = 1, \qquad x_3(0) = -1.$$

18. $\dot{x}_1 = -6x_3,$
$\dot{x}_2 = x_1 - x_3,$
$\dot{x}_3 = x_2 + 4x_3.$

Also, find the solution that satisfies the conditions
$$x_1(0) = 1, \qquad x_2(0) = -1, \qquad x_3(0) = 1.$$

19. $\dot{x}_1 = x_2 + 5x_3 + 9x_4,$
$\dot{x}_2 = 2x_1 + x_2 + 6x_3 + 8x_4,$
$\dot{x}_3 = x_4,$
$\dot{x}_4 = x_3 - 2x_4.$

20. $\dot{x}_1 = -x_4,$
$\dot{x}_2 = x_1 - 4x_4,$
$\dot{x}_3 = x_2 - 6x_4,$
$\dot{x}_4 = x_3 - 4x_4.$

21. $\dot{x}_1 = -3x_1 + 3x_2 - x_3,$
$\dot{x}_2 = -2x_1 + x_2,$
$\dot{x}_3 = -x_1 + x_2 - x_3.$

22. Refer to Exercise 1 and solve the system
$$\dot{x} = x_2 + t,$$
$$\dot{x}_2 = -2x_1 + 3x_2 - 1.$$

23. Refer to Exercise 2 and solve the system
$$\dot{x}_1 = -3x_1 + 4x_2 + \sin t,$$
$$\dot{x}_2 = -2x_1 + 3x_2 - 2.$$

24. Refer to Exercise 3 and solve the system
$$\dot{x}_1 = x_2 + 2t,$$
$$\dot{x}_2 = -x_1 + 2x_2 - 3.$$

25.   Refer to Exercise 4 and solve the system

$$\dot{x}_1 = 4x_1 - x_2 - 3e^{2t},$$
$$\dot{x}_2 = 4x_1 + 1.$$

Answers

1. $c_1e^t + c_2e^{2t},$
    $c_1e^t + 2c_2e^{2t}.$

2. $c_1e^t + 2c_2e^{-t},$
    $c_1e^t + c_2e^{-t}.$

3. $c_1e^t + c_2te^t,$
    $c_1e^t + c_2(t + 1)e^t.$

4. $c_1e^{2t} + c_2te^{2t},$
    $2c_1e^{2t} + c_2(2t - 1)e^{2t}.$

5. $c_1e^{2t} + c_2e^{-3t},$
    $2c_1e^{2t} + 3c_2e^{-3t}.$

6. $2c_1e^t \cos t + 2c_2e^t \sin t,$
    $c_1e^t(\sin t - \cos t) - c_2e^t(\sin t + \cos t).$

7. $c_1e^{3t} \cos 2t + c_2e^{3t} \sin 2t,$
    $c_1e^{3t}(\sin 2t - \cos 2t) - c_2e^{3t}(\sin 2t + \cos 2t).$

8. $c_1e^{2t} \cos 2t + c_2e^{2t} \sin 2t,$
    $c_1e^{2t}(\cos 2t + \sin 2t) + c_2e^{2t}(\sin 2t - \cos 2t).$

9. $\begin{array}{ccc} e^t & e^{-t} & e^{2t} \\ e^t & e^{-t} & 2e^{2t} \\ e^t & 2e^{-t} & e^{2t} \end{array}.$

10. $\begin{array}{ccc} e^t & te^t & e^{2t} \\ e^t & (t + 1)e^t & 2e^{2t} \\ e^t & (t + 2)e^t & e^{2t}. \end{array}$

11. $\begin{array}{ccc} e^t & e^t(2 \cos t - \sin t) & e^t(\cos t + 2 \sin t) \\ e^t & 2e^t \cos t & 2e^t \sin t \\ e^t & e^t \cos t & e^t \sin t. \end{array}$

12. $\begin{array}{ccc} e^t & e^{-2t} & te^{-2t} \\ e^t & -2e^{-2t} & (-2t + 1)e^{-2t} \\ e^t & 4e^{-2t} & (4t - 4)e^{-2t}. \end{array}$

13. $\begin{array}{ccc} e^{2t} & te^{2t} & t^2e^{2t} \\ 2e^{2t} & (2t + 1)e^{2t} & (2t^2 + 2t)e^{2t} \\ 4e^{2t} & (4t + 4)e^{2t} & (4t^2 + 8t + 2)e^{2t}. \end{array}$

15. $9e^t$ $5e^{-t}$ $4e^{2t}$ $-9e^t + 5e^{-t} + 4e^{2t}$
$3e^t$ $2e^{-t}$ $2e^{2t}$ $-3e^t + e^{-t} + 2e^{2t}$
$2e^t$ $2e^{-t}$ $e^{2t};$ $-2e^t + 2e^{-t} + e^{2t}.$

17. e^t $2e^{4t}$ $2e^{16t}$ $\frac{1}{3}(2e^{4t} - 2e^{16t})$
$2e^t$ e^{4t} $-2e^{16t}$ $\frac{1}{3}(e^{4t} + 2e^{16t})$
$2e^t$ $-2e^{4t}$ $e^{16t};$ $\frac{1}{3}(-2e^{4t} - e^{16t}).$

19. $-13e^t$ e^{-t} e^{2t} $-7e^{-3t}$
$-37e^t$ $-e^{-t}$ $2e^{2t}$ e^{-3t}
$3e^t$ 0 0 $-5e^{-3t}$
e^t 0 0 $5e^{-3t}.$

21. e^{-t} $e^{-t}(2\cos t + \sin t)$ $e^{-t}(2\sin t - \cos t)$
e^{-t} $2e^{-t}\cos t$ $2e^{-t}\sin t$
e^{-t} $e^{-t}\cos t$ $e^{-t}\sin t.$

22. $-\frac{3}{2}t - 1 + c_1 e^t + c_2 e^{2t}$
$-t - \frac{3}{2} + c_1 e^t + 2c_2 e^{2t}.$

23. A particular solution is

$$\tfrac{3}{2}\sin t - \tfrac{1}{2}\cos t + 8$$
$$\sin t + 6.$$

24. A particular solution is

$$-4t - 9$$
$$-2t - 4.$$

25. A particular solution is

$$-3t^2 e^{2t} - 3t e^{2t} - \tfrac{1}{4}$$
$$-6t^2 e^{2t} - 1.$$

4 Matrix notation

We may consider the column

$$x = \begin{bmatrix} x_1 \\ x_2 \\ \vdots \\ x_n \end{bmatrix}$$

as an n-dimensional column vector, the kth component of which is x_k. It may also be regarded as an $n \times 1$ matrix—that is, a matrix with n rows and 1 column. If the components are functions of t on an interval I, the column becomes a *vector-valued* function on I, since each value of t on the interval determines a vector x. The vector-valued function is said to be continuous on I if each component is continuous on I.

The sum of two column vectors x and y is the column vector

$$x + y = \begin{bmatrix} x_1 + y_1 \\ x_2 + y_2 \\ \vdots \\ x_n + y_n \end{bmatrix},$$

where y_k $(k = 1, 2, \ldots, n)$ is, of course, the kth component of y. If c is a constant—that is, a *scalar*—the *product*

$$cx = \begin{bmatrix} cx_1 \\ cx_2 \\ \vdots \\ cx_n \end{bmatrix}.$$

Similarly, the *derivative* with respect to t of the vector x is defined to be the vector

$$\dot{x} = \begin{bmatrix} \dot{x}_1 \\ \dot{x}_2 \\ \vdots \\ \dot{x}_n \end{bmatrix},$$

while the *integral*

$$\int_a^b x(t)\, dt = \begin{bmatrix} \displaystyle\int_a^b x_1(t)\, dt \\ \displaystyle\int_a^b x_2(t)\, dt \\ \vdots \\ \displaystyle\int_a^b x_n(t)\, dt \end{bmatrix}.$$

Two vectors x and y are said to be *equal* if and only if $x_k = y_k$ $(k = 1, 2, \ldots, n)$. The *null vector* 0 is the vector each of whose components is zero. The *norm* $\|x\|$ of a vector x may be defined in a number of ways. The usual definition is

$$\|x\| = (x_1^2 + x_2^2 + \cdots + x_n^2)^{1/2}.$$

If we designate by $A(t)$ the matrix

$$A(t) = \begin{bmatrix} a_{11}(t) & a_{12}(t) & \cdots & a_{1n}(t) \\ a_{21}(t) & a_{22}(t) & \cdots & a_{2n}(t) \\ \cdots\cdots\cdots\cdots\cdots\cdots\cdots \\ a_{n1}(t) & a_{n2}(t) & \cdots & a_{nn}(t) \end{bmatrix},$$

the *product* $A(t)x$ of the matrices $A(t)$ and x is, as we shall see in a moment, precisely the right-hand member of (1.1), and the system (1.1) can then be written

(4.1)
$$\frac{dx}{dt} = A(t)x.$$

We say that $A(t)$ is a *continuous* matrix if each element $a_{ij}(t)$ is continuous. A *solution* of (4.1) is then a vector-valued function $x(t)$—that is, a column such as (1.2).

Theorem 1.1 can now be translated as follows:

Suppose $A(t)$ is a continuous matrix on an interval I, let t_0 be any point of I, and let

$$a = \begin{bmatrix} a_1 \\ a_2 \\ \vdots \\ a_n \end{bmatrix}$$

be an arbitrary constant vector. There then exists one and only one solution $x(t)$ of the system

$$\frac{dx}{dt} = A(t)x,$$

$$x(t_0) = a.$$

This solution is valid on the entire interval I.

The value of matrix notation does not, of course, lie solely in its shorthand characteristics. It also permits some of the methods of linear algebra to be employed in dealing with systems of linear differential equations. Students unfamiliar with matrix manipulation may find the following observations helpful. An $m \times n$ matrix is a rectangular array of numbers

$$\begin{bmatrix} a_{11} & a_{12} & \cdots & a_{1n} \\ a_{21} & a_{22} & \cdots & a_{2n} \\ \cdots\cdots\cdots\cdots\cdots\cdots \\ a_{m1} & a_{m2} & \cdots & a_{mn} \end{bmatrix}$$

of m rows and n columns. The *dimensions* of the matrix are $m \times n$. The numbers a_{ij} are called *elements* of the matrix.

Two matrices are said to be *equal* if corresponding elements are equal. To be equal, they must then have the same dimensions. A *null* matrix is one all of whose elements are zero.

The sum of two matrices of the same dimensions is simply the matrix obtained by adding corresponding elements. Thus, the sum of two 2×3 matrices is given by the equation

$$\begin{bmatrix} a_{11} & a_{12} & a_{13} \\ a_{21} & a_{22} & a_{23} \end{bmatrix} + \begin{bmatrix} b_{11} & b_{12} & b_{13} \\ b_{21} & b_{22} & b_{23} \end{bmatrix} = \begin{bmatrix} a_{11} + b_{11} & a_{12} + b_{12} & a_{13} + b_{13} \\ a_{21} + b_{21} & a_{22} + b_{22} & a_{23} + b_{23} \end{bmatrix}.$$

Multiplying a matrix by a constant k means multiplying each element by k.

If the elements of a matrix are functions of a variable t, the *derivative* of the matrix is the matrix of the derivatives of the individual elements. Thus, if

$$A = \begin{bmatrix} t & t^2 \\ \sin t & 1 \end{bmatrix},$$

then

$$\frac{dA}{dt} = \begin{bmatrix} 1 & 2t \\ \cos t & 0 \end{bmatrix}.$$

A similar statement holds for *integration* of a matrix; accordingly, if A is the matrix in the above example,

$$\int_0^\pi A \, dt = \begin{bmatrix} \int_0^\pi t \, dt & \int_0^\pi t^2 \, dt \\ \int_0^\pi \sin t \, dt & \int_0^\pi dt \end{bmatrix} = \begin{bmatrix} \dfrac{\pi^2}{2} & \dfrac{\pi^3}{3} \\ 2 & \pi \end{bmatrix}.$$

These definitions will be seen to be simply extensions of the corresponding definitions for operations with vectors.

Matrix multiplication. The *product* of two matrices AB is defined only when A is an $m \times r$ matrix and B is an $r \times n$ matrix—that is, when the number of columns in A is the same as the number of rows in B. The resulting matrix is $m \times n$, the elements of which are obtained as illustrated below.

$$\begin{bmatrix} * & * & * & * \\ a & b & c & d \\ * & * & * & * \end{bmatrix} \begin{bmatrix} * & * & A \\ * & * & B \\ * & * & C \\ * & * & D \end{bmatrix} = \begin{bmatrix} * & * & & * \\ * & * & (aA + bB + cC + dD) \\ * & * & & * \end{bmatrix}.$$

Here, A is a 3×4 matrix, while B is 4×3. In the product matrix we have shown only the element in the second row and third column. The other elements in the product are obtained in similar fashion. Thus,

$$(4.2) \qquad \begin{bmatrix} a_{11} & a_{12} & a_{13} \\ a_{21} & a_{22} & a_{23} \\ a_{31} & a_{32} & a_{33} \end{bmatrix} \begin{bmatrix} x \\ y \\ z \end{bmatrix} = \begin{bmatrix} a_{11}x + a_{12}y + a_{13}z \\ a_{21}x + a_{22}y + a_{23}z \\ a_{31}x + a_{32}y + a_{33}z \end{bmatrix}.$$

The meaning of equation (4.1) is clear from (4.2).

Matrix multiplication is not *commutative*; that is, AB is not, in general, equal to BA. Indeed, AB may be defined when BA is not. In (4.2), for example, the expression

$$\begin{bmatrix} x \\ y \\ z \end{bmatrix} A$$

has no meaning, for the first matrix has one column while the second has three rows. But even when both matrices are square and have the same dimensions, multiplication is not commutative, as the following example shows:

$$\begin{bmatrix} 1 & 1 \\ 0 & 1 \end{bmatrix} \begin{bmatrix} 1 & 1 \\ 1 & 1 \end{bmatrix} = \begin{bmatrix} 2 & 2 \\ 1 & 1 \end{bmatrix}, \qquad \begin{bmatrix} 1 & 1 \\ 1 & 1 \end{bmatrix} \begin{bmatrix} 1 & 1 \\ 0 & 1 \end{bmatrix} = \begin{bmatrix} 1 & 2 \\ 1 & 2 \end{bmatrix}.$$

Matrix multiplication is, however, *associative*; that is, if A, B, C are matrices for which the product $A(BC)$ is defined, then

$$(4.3) \qquad\qquad A(BC) = (AB)C.$$

The proof of this is not difficult, but it belongs in linear algebra and is omitted. Because of (4.3) one frequently expresses either product in that equation by the symbol ABC.

To illustrate the associativity of multiplication, consider the products

$$\begin{bmatrix} 1 & 1 \\ 0 & 1 \end{bmatrix} \left(\begin{bmatrix} 1 & 1 \\ 1 & -1 \end{bmatrix} \begin{bmatrix} x \\ y \end{bmatrix} \right) = \begin{bmatrix} 1 & 1 \\ 0 & 1 \end{bmatrix} \begin{bmatrix} x + y \\ x - y \end{bmatrix} = \begin{bmatrix} 2x \\ x - y \end{bmatrix},$$

$$\left(\begin{bmatrix} 1 & 1 \\ 0 & 1 \end{bmatrix} \begin{bmatrix} 1 & 1 \\ 1 & -1 \end{bmatrix} \right) \begin{bmatrix} x \\ y \end{bmatrix} = \begin{bmatrix} 2 & 0 \\ 1 & -1 \end{bmatrix} \begin{bmatrix} x \\ y \end{bmatrix} = \begin{bmatrix} 2x \\ x - y \end{bmatrix}.$$

The *identity matrix E* is an $n \times n$ matrix of the form

$$E = \begin{bmatrix} 1 & 0 & \cdots & 0 \\ 0 & 1 & \cdots & 0 \\ \hdotsfor{4} \\ 0 & 0 & \cdots & 1 \end{bmatrix};$$

that is, a square matrix with elements δ_{ij}, where $\delta_{ii} = 1$, $\delta_{ij} = 0$ $(i \neq j)$. This quantity δ_{ij} is called the *Kronecker delta*. If A is an arbitrary $n \times n$ matrix, it is easy to see that

$$EA = AE = A.$$

Multiplication of a matrix by a scalar, say k, may be regarded as multiplication by $kE = Ek$ and is also easily seen to be commutative.

Consider now the differential equation

(4.4)
$$\frac{dx}{dt} = Ax,$$

where A is an $n \times n$ matrix of constants, with a_{ij} the element in its ith row and jth column, and x is a column vector of n elements† x_i. The substitution

$$x = Be^{\lambda t}$$

in (4.4), where B is a column vector, leads consecutively to the matrix equations

$$B\lambda e^{\lambda t} = A Be^{\lambda t},$$

(4.5)
$$0 = e^{\lambda t}(A - \lambda E)B,$$

$$(A - \lambda E)B = 0.$$

In order that a nonnull vector B exist satisfying the last equation (4.5), it is necessary and sufficient that the determinant

$$|A - \lambda E| = 0,$$

but this is simply the familiar characteristic equation

(4.6)
$$\begin{vmatrix} (a_{11} - \lambda) & a_{12} & \cdots & a_{1n} \\ a_{21} & (a_{22} - \lambda) & \cdots & a_{2n} \\ \cdots\cdots\cdots\cdots\cdots\cdots\cdots\cdots\cdots\cdots\cdots \\ a_{n1} & a_{n2} & \cdots & (a_{nn} - \lambda) \end{vmatrix} = 0.$$

Example. Solve the equation [see (3.9)]

(4.7)
$$\frac{dx}{dt} = Ax,$$

where

$$A = \begin{bmatrix} \frac{1}{2} & \frac{1}{2} \\ -\frac{3}{2} & \frac{5}{2} \end{bmatrix}.$$

---

† Or, "components." When x is regarded primarily as a vector, it is customary to speak of its "components"—when it is primarily regarded as an $n \times 1$ matrix, it is customary to speak of its "elements."

The characteristic equation is

$$\begin{vmatrix} \frac{1}{2} - \lambda & \frac{1}{2} \\ -\frac{3}{2} & \frac{5}{2} - \lambda \end{vmatrix} = \lambda^2 - 3\lambda + 2 = 0,$$

the roots of which are $\lambda = 1, 2$. The last equation (4.5) provides a vector B corresponding to each of these characteristic roots, and the solution of (4.7) proceeds essentially as in the earlier treatment of this example. One would, of course, write the answer in vector form; that is, the general solution of (4.7) is the vector [see (3.13)]

$$(4.8) \qquad \begin{bmatrix} c_1 e^t + c_2 e^{2t} \\ c_1 e^t + 3c_2 e^{2t} \end{bmatrix}.$$

If we were seeking the solution $x(t)$ of this vector differential equation with the property that

$$x(0) = \begin{bmatrix} 0 \\ 1 \end{bmatrix},$$

we would set $t = 0$ in (4.8) and have

$$c_1 + c_2 = 0,$$

$$c_1 + 3c_2 = 1.$$

It would follow that $c_1 = -\frac{1}{2}$, $c_2 = \frac{1}{2}$, and

$$x(t) = \frac{1}{2} \begin{bmatrix} -e^t + e^{2t} \\ -e^t + 3e^{2t} \end{bmatrix}.$$

The *transpose* of a matrix is an important concept. It is the matrix obtained from a given matrix by interchanging its rows and columns. Various symbols are used in the literature for the transpose of a matrix. We shall use an asterisk *. Thus, if

$$A = \begin{bmatrix} a_{11} & a_{12} & a_{13} \\ a_{21} & a_{22} & a_{23} \end{bmatrix},$$

$$A^* = \begin{bmatrix} a_{11} & a_{21} \\ a_{12} & a_{22} \\ a_{13} & a_{23} \end{bmatrix}.$$

It is a theorem in linear algebra that

$$(AB)^* = B^*A^*.$$

The product

$$\begin{bmatrix} a_{11} & a_{12} \\ a_{21} & a_{22} \end{bmatrix} \begin{bmatrix} x_1 \\ x_2 \end{bmatrix}$$

can then, for example, be written as

$$\begin{bmatrix} a_{11} & a_{12} \\ a_{21} & a_{22} \end{bmatrix} [x_1 \quad x_2]^*,$$

and so on.

If the matrices A and B are functions of t that possess derivatives, a frequently useful formula (that we shall not prove) is the following:

$$\frac{d}{dt}(AB) = \frac{dA}{dt}B + A\frac{dB}{dt}.$$

That is to say that differentiation of a product of matrices follows the same rule as differentiation of the product of two scalar functions.

It is occasionally useful to write a second-order linear differential equation

$$\ddot{x} + b(t)\dot{x} + c(t)x = 0$$

as a pair of first-order equations by means of the substitution $\dot{x} = y$. We would then have

$$\dot{x} = y,$$
$$\dot{y} = -c(t)x - b(t)y.$$

Consider, for example, the differential equation

(4.9) $\ddot{x} + \dot{x} - 2x = 0.$

Using the substitution $\dot{x} = y$ we have the pair of equations

(4.10)
$$\dot{x} = y,$$
$$\dot{y} = 2x - y.$$

The characteristic equation becomes

$$\begin{vmatrix} -\lambda & 1 \\ 2 & -1 - \lambda \end{vmatrix} = \lambda^2 + \lambda - 2 = 0,$$

the roots of which are, as is to be expected, $\lambda = 1, -2$. The last equation (4.5) becomes

$$-\lambda B_1 + B_2 = 0$$
$$2B_1 - (1 + \lambda)B_2 = 0.$$

If we set $\lambda = 1$, the vector (B_1, B_2) can then be taken as the vector $(1, 1)$, while the root $\lambda = -2$ yields the vector $(B_1, B_2) = (1, -2)$. A fundamental system of solutions of (4.10) is then

$$\begin{bmatrix} e^t & e^{-2t} \\ e^t & -2e^{-2t} \end{bmatrix},$$

or the general solution can be written as

$$x = c_1 e^t + c_2 e^{-2t}$$
$$y = c_1 e^t - 2c_2 e^{-2t}.$$

Note that the general solution of (4.9) is given by the first equation above, while the second equation yields its derivative.

The material in this section has provided an opportunity to list some of the elements of matrix manipulation. Systematic treatments of linear algebra are given in a large number of books.†

Exercises

1. Write Exercise 1 of Section 3 in vector form and solve.

2. Write Exercise 2 of Section 3 in vector form and solve.

3. Write Exercise 3 of Section 3 in vector form and solve.

4. Write Exercise 4 of Section 3 in vector form and solve.

5. Write Exercise 5 of Section 3 in vector form and solve.

6. Write Exercise 9 of Section 3 in vector form and solve.

7. Write Exercise 10 of Section 3 in vector form and solve.

8. Solve the matrix differential system

$$\frac{dx}{dt} = A(t)x$$

when

(a) $A(t) = \begin{pmatrix} \frac{1}{2} & \frac{1}{2} \\ -\frac{3}{2} & \frac{5}{2} \end{pmatrix},$ $\qquad x(0) = \begin{bmatrix} 1 \\ 1 \end{bmatrix};$

(b) $A(t) = \begin{pmatrix} 0 & 1 \\ -2 & 3 \end{pmatrix},$ $\qquad x(0) = \begin{bmatrix} 0 \\ 1 \end{bmatrix};$

(c) $A(t) = \begin{pmatrix} 0 & 2 & 0 \\ -1 & 3 & 0 \\ 0 & 0 & 3 \end{pmatrix},$ $\qquad x(0) = \begin{bmatrix} 0 \\ 1 \\ 1 \end{bmatrix}.$

---

† One attractive treatment is to be found in H. W. Brinkmann and E. A. Klotz, *Linear Algebra and Analytic Geometry*, Addison-Wesley Publishing Co., Reading, Massachusetts (1971).

9. Given the system

$$\frac{dx}{dt} = \begin{bmatrix} a & b \\ c & d \end{bmatrix} \begin{bmatrix} x_1 \\ x_2 \end{bmatrix},$$

where the a_{ij} are constants, show that

$$\ddot{x}_1 - (a + d)\dot{x}_1 + (ad - bc)x_1 = 0.$$

10. Verify that $(AB)^* = B^*A^*$ when

$$A = \begin{pmatrix} a & b \\ c & d \end{pmatrix}, \qquad B = \begin{pmatrix} e & f \\ g & h \end{pmatrix}.$$

11. Verify that

$$(AB)^{\cdot} = \dot{A}B + A\dot{B}$$

when

$$A = \begin{pmatrix} t & 1 \\ 0 & t^2 \end{pmatrix}, \qquad B = \begin{pmatrix} \sin t & 1 \\ \cos t & 0 \end{pmatrix}.$$

12. The *inverse* A^{-1} of a matrix A, when the inverse exists, is a matrix such that

$$AA^{-1} = E.$$

Find the inverse of the matrix

$$A = \begin{bmatrix} 1 & -1 \\ 2 & 0 \end{bmatrix}.$$

Show that $A^{-1}A = E$ also.

13. Refer to Exercise 12 and find A^{-1} when

$$A = \begin{bmatrix} a & b \\ c & d \end{bmatrix},$$

and

$$\begin{vmatrix} a & b \\ c & d \end{vmatrix} \neq 0.$$

14. An important matrix equation that arises in stability theory is the equation

$$A^*B + BA = -8E,$$

where A is given and B is an unknown matrix. Solve this equation for B when

$$A = \begin{bmatrix} 1 & 1 \\ 1 & -3 \end{bmatrix}.$$

Answers

8. (a) $x = \begin{bmatrix} e^t \\ e^t \end{bmatrix}$;

(b) $x = \begin{bmatrix} e^{2t} & -e^t \\ 2e^{2t} & -e^t \end{bmatrix}$;

(c) $x = \begin{bmatrix} 2e^{2t} & -2e^t \\ 2e^{2t} & -e^t \\ & e^{3t} \end{bmatrix}$

13. $\begin{bmatrix} d & -b \\ -c & a \end{bmatrix} \dfrac{1}{ad - bc}$.

14. $\begin{bmatrix} -3 & -1 \\ -1 & 1 \end{bmatrix}$.

5 Laplace transforms for systems

The methods of Laplace transforms are available for some simultaneous linear differential equations. The procedure to be followed will be clear from the following examples.

Example. Find a pair of functions $x(t)$, $y(t)$ that satisfy the simultaneous differential equations

(5.1)
$$\dot{x} = -x - 5y, \qquad x(0) = 1,$$
$$2\dot{x} - \dot{y} = -12y, \qquad y(0) = 1.$$

Taking the Laplace transform of both sides of both equations and employing equation (2.3) of Chapter 4 we have

$$(s + 1)\mathscr{L}[x] + 5\mathscr{L}[y] = 1,$$
$$2s\mathscr{L}[x] + (12 - s)\mathscr{L}[y] = 1.$$

These are simultaneous equations in $\mathscr{L}[x]$ and $\mathscr{L}[y]$. Solving them we have

$$\mathscr{L}[x] = \frac{s - 7}{(s - 4)(s + 3)} = \frac{-\frac{3}{7}}{s - 4} + \frac{\frac{10}{7}}{s + 3},$$

$$\mathscr{L}[y] = \frac{s - 1}{(s - 4)(s + 3)} + \frac{\frac{3}{7}}{s - 4} + \frac{\frac{4}{7}}{s + 3} \qquad (s > 4).$$

Accordingly,

(5.2)
$$x(t) = -\tfrac{3}{7}e^{4t} + \tfrac{10}{7}e^{-3t},$$
$$y(t) = \tfrac{3}{7}e^{4t} + \tfrac{4}{7}e^{-3t}.$$

Substitution of (5.2) in (5.1) provides a check on the accuracy of our computation.

Example.   Find a pair of functions $x(t)$, $y(t)$ that satisfy the simultaneous differential equations

(5.3)
$$\dot{x} + 2x + y = 2t, \qquad x(0) = 0, \qquad y(0) = 0,$$
$$\ddot{y} + 2\dot{y} + \dot{x} = 1, \qquad y'(0) = 1.$$

Taking the Laplace transform of both sides of both equations and using equations (2.3) and (2.4) of Chapter 4 we have the simultaneous equations

$$(s + 2)\mathscr{L}[x] + \mathscr{L}[y] = \frac{2}{s^2},$$

$$s\mathscr{L}[x] + (s^2 + 2s)\mathscr{L}[y] = \frac{s + 1}{s}$$

for the determination of $\mathscr{L}[x]$ and $\mathscr{L}[y]$. Solving these equations we have

$$\mathscr{L}[x] = \frac{1}{s^2(s + 1)} = -\frac{1}{s} + \frac{1}{s^2} + \frac{1}{s + 1},$$

$$\mathscr{L}[y] = \frac{1}{s(s + 1)} = \frac{1}{s} - \frac{1}{s + 1}.$$

It follows that

(5.4)
$$x(t) = -1 + t + e^{-t},$$
$$y(t) = 1 - e^{-t}.$$

Again, the substitution of (5.4) in (5.3) provides a check on the accuracy of computation.

Exercises

Solve the given simultaneous systems of differential equations.

1. $\dot{x} = y$, $\dot{y} = -2x + 3y$, $x(0) = 0$, $y(0) = 1$.

2. $\dot{x} = -3x + 4y + \sin t$, $\dot{y} = -2x + 3y - 2$, $x(0) = 3$, $y(0) = 4$.

3. $\dot{x} = y + 2t$, $\dot{y} = -x + 2y - 3$, $x(0) = 1$, $y(0) = 0$.

4. $\dot{x} = y + \sin t$, $\dot{y} = -2x + 3y - \cos t$, $x(0) = 1$, $y(0) = 1$.

5. $\ddot{x} + 3\dot{x} - 4\dot{y} = \cos t$, $\dot{y} + 2x - 3y = -2$, $x(0) = 0$, $\dot{x}(0) = 1$, $y(0) = 1$.

6. $\ddot{x} - \dot{y} = 4t + 1$, $\ddot{y} + 2\dot{x} - 3\dot{y} = 1$, $x(0) = \dot{x}(0) = 0$, $y(0) = 0$, $\dot{y}(0) = 1$.

7. $\dot{x} - 2x - y + 2z = 0$,
$\dot{y} - 3y + 2z = 0$,
$\dot{z} - 3x - y + 3z = 0$,
$x(0) = 1$, $y(0) = 2$, $z(0) = 2$.

Answers

1. $x(t) = e^{2t} - e^t$,
$y(t) = 2e^{2t} - e^t$.

2. $x(t) = \frac{3}{2} \sin t - \frac{1}{2} \cos t + 8 + \frac{1}{2}e^t - 5e^{-t}$,
$y(t) = \sin t + 6 + \frac{1}{2}e^t - \frac{5}{2}e^{-t}$.

3. $x(t) = -4t - 9 + 10e^t - 6te^t$.
$y(t) = -2t - 4 + 4e^t - 6te^t$.

4. $x(t) = -\frac{3}{10} \sin t - \frac{9}{10} \cos t + \frac{5}{2}e^t - \frac{3}{5}e^{2t}$,
$y(t) = -\frac{1}{10} \sin t - \frac{3}{10} \cos t + \frac{5}{2}e^t - \frac{6}{5}e^{2t}$.

5. $x(t) = -1 + \frac{1}{2}e^t + e^{-t} - \frac{1}{2} \cos t + \frac{3}{2} \sin t$,
$y(t) = \frac{1}{2}e^t + \frac{1}{2}e^{-t} + \sin t$.

6. $x(t) = -8 - 8t - 3t^2 + 8e^t$,
$y(t) = -8 - 7t - 2t^2 + 8e^t$.

7. $x(t) = e^{-t} - e^t + e^{2t}$,
$y(t) = e^{-t} - e^t + 2e^{2t}$,
$z(t) = 2e^{-t} - e^t + e^{2t}$.

10

Autonomous systems in the plane

1 The Poincaré phase plane

In a system of differential equations such as

$$\frac{dx_1}{dt} = f_1(x_1, x_2, t),$$

$$\frac{dx_2}{dt} = f_2(x_1, x_2, t)$$

the right-hand members may not depend formally on the time t. The equations then have the form

$$\frac{dx_1}{dt} = f_1(x_1, x_2),$$

$$\frac{dx_2}{dt} = f_2(x_1, x_2).$$

Such a system is said to be *autonomous* or *time-invariant*.

In this chapter we shall be principally concerned with autonomous linear systems

(1.1)
$$\frac{dx}{dt} = ax + by,$$

$$\frac{dy}{dt} = cx + dy \qquad (ad - bc \neq 0),$$

where a, b, c, and d are constants and t is time. Such systems are of fundamental importance in the study of second-order autonomous systems that are not necessarily linear. As we have seen in Chapter 9, a solution of a system (1.1) may be written as a column

$$x = x(t)$$
$$y = y(t).$$

If, now, t is regarded as a parameter, this pair of equations will represent a curve in the xy-plane. Such a curve is called a *trajectory* of the system (1.1), and the xy-plane is then said to be the *phase plane* of this system.

Example. The system

$$\frac{dx}{dt} = y,$$

$$\frac{dy}{dt} = -x$$

has the general solution

$$x = c_1 \sin t + c_2 \cos t$$
$$y = c_1 \cos t - c_2 \sin t.$$

In this example it is easy to see that the trajectories are circles in the phase plane, for

$$x^2 + y^2 = k^2 \qquad (k^2 = c_1^2 + c_2^2).$$

This result could also have been obtained by observing that the variable t can be eliminated by writing

$$\frac{dy}{dx} = -\frac{x}{y};$$

accordingly,

$$x \, dx + y \, dy = 0$$

the integral curves (trajectories) of which are

$$x^2 + y^2 = k^2.$$

A point at which both dx/dt and dy/dt vanish—where

$$\left(\frac{dx}{dt}\right)^2 + \left(\frac{dy}{dt}\right)^2 = 0$$

—is called an *equilibrium point* of the system (1.1). The origin is such a point. Note that if the determinant

$$\begin{vmatrix} a & b \\ c & d \end{vmatrix} = ad - bc$$

is not equal to 0, the origin is the only equilibrium point of the system (1.1).

We seek a solution of (1.1) in the form

(1.2)
$$x = Ae^{\lambda t},$$
$$y = Be^{\lambda t},$$

where A, B, and λ are suitably chosen constants (see Chapter 9). When the substitution (1.2) is made in equations (1.1), we have

(1.3)
$$A(a - \lambda) + Bb = 0,$$
$$Ac + B(d - \lambda) = 0.$$

Equations (1.3) will possess solutions A and B not both zero if and only if λ is a root of the characteristic equation

(1.4)
$$\begin{vmatrix} a - \lambda & b \\ c & d - \lambda \end{vmatrix} = 0,$$

or,

(1.5)
$$\lambda^2 - (a + d)\lambda + (ad - bc) = 0.$$

If equation (1.5) is written as

(1.6)
$$\lambda^2 - p\lambda + q = 0,$$

where

$$p = a + d,$$
$$q = ad - bc,$$

we note that

$$p = \lambda_1 + \lambda_2,$$

$$q = \lambda_1 \lambda_2.$$

where λ_1 and λ_2 are the characteristic roots. The discriminant Δ of equation (1.6) is given by

$$\Delta = p^2 - 4q$$

$$= (a - d)^2 + 4bc.$$

The behavior of the solutions of equations (1.1) depends in a fundamental way on the nature of the roots of equation (1.6). Let us examine these solutions more carefully.

If λ is set equal to λ_1 in the first of equations (1.3) and if we designate by A_1 and B_1, respectively, any corresponding nontrivial solution A and B of (1.3), we have

$$A_1(a - \lambda_1) + B_1 b = 0.$$

Similarly,

$$A_2(a - \lambda_2) + B_2 b = 0,$$

where A_2 and B_2 are any nontrivial solution A and B of (1.3) corresponding to the characteristic root λ_2.

If $\lambda_1 \neq \lambda_2$, linearly independent solutions of (1.1) are furnished by the columns

(1.7)
$$\begin{matrix} A_1 e^{\lambda_1 t} & A_2 e^{\lambda_2 t} \\ B_1 e^{\lambda_1 t} & B_2 e^{\lambda_2 t}. \end{matrix}$$

The general solution of (1.1) may then be written in the form

(1.8)
$$x = c_1 A_1 e^{\lambda_1 t} + c_2 A_2 e^{\lambda_2 t},$$

$$y = c_1 B_1 e^{\lambda_1 t} + c_2 B_2 e^{\lambda_2 t},$$

where c_1 and c_2 are arbitrary constants.

The trajectories determined by (1.8), when t is regarded as a parameter, satisfy at each point the differential equation

(1.9)
$$(cx + dy)\, dx = (ax + by)\, dy.$$

We observe that, by the fundamental existence theorem, there is one and only one of these curves through each point of the plane, except possibly the origin.†

---

† At a point, not the origin, on the line $ax + by = 0$ note that $cx + dy \neq 0$, and the existence theorem may then be applied to the differential equation

$$\frac{dx}{dy} = \frac{ax + by}{cx + dy}.$$

A number of cases are illustrated below by examples. An examination of the examples will indicate that the methods employed can be applied in general.

Case 1: λ_1, λ_2 *real and of opposite signs.* In this case $q < 0$. As an example, consider the system

(1.10)
$$\frac{dx}{dt} = x - y,$$
$$\frac{dy}{dt} = -2x.$$

The characteristic equation is

$$\begin{vmatrix} 1 - \lambda & -1 \\ -2 & -\lambda \end{vmatrix} = \lambda^2 - \lambda - 2 = 0,$$

and the characteristic roots are $\lambda_1 = -1$, $\lambda_2 = 2$. The general solution may be written in the form

(1.11)
$$x = c_1 e^{-t} + c_2 e^{2t},$$
$$y = 2c_1 e^{-t} - c_2 e^{2t}.$$

Consider first the case when $c_1 = c_2 = 0$. The solution (1.11) becomes the null solution represented in the phase plane by the point at the origin. Next, suppose that $c_2 = 0$ and $c_1 \neq 0$. The points (x, y) of a solution in the phase plane lie on two rays—the line $y = 2x$ with the origin deleted. As t varies from an arbitrary fixed value t_0 to $+\infty$, the point (x, y) moves along its ray toward the origin as a limit. Similarly, if $c_1 = 0$, $c_2 \neq 0$, the point (x, y) moves along the line $y = -x$ (with the origin deleted) away from the origin, as t varies from a fixed value t_0 to $+\infty$ (see Fig. 10.1).

Either by solving the differential equation (1.9) or by eliminating the parameter t in equation (1.11) we find that the curves represented by (1.11) have the equation

(1.12) $(x + y)^2(y - 2x) = k,$

where k is a constant. It is clear from equation (1.12) that the lines $x + y = 0$ and $y - 2x = 0$ are asymptotes of the family (1.12). This can also be observed from equations (1.11). For, when t is very large and positive, the first terms in the right-hand members of (1.11) are very small, and these equations are approximately $x = c_2 e^{2t}$, $y = -c_2 e^{2t}$; that is, the point (x, y) is close to the line $y = -x$. Similarly, when t is numerically large but negative, the point (x, y) is near the line $y = 2x$.

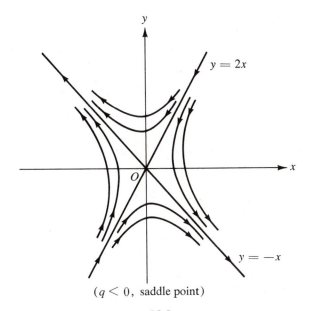

$(q < 0,$ saddle point$)$

FIG. 10.1

Note that the curves (1.11) have slopes given by the formula

$$\frac{dy}{dx} = \frac{2x}{y - x}.$$

This follows from (1.10). Thus, the *trajectories*—the curves (1.11)—have the property that all have horizontal tangents where they cross the y-axis. Further, they have vertical tangents at the points where they cross the line $y = x$, and all have slope -2 at their x-intercepts. It is clear from (1.11) that if neither c_1 nor c_2 is zero, a point (x, y) moving along a trajectory cannot approach the origin.

Typical trajectories in the phase plane are shown in Fig. 10.1. The arrowheads on the curves indicate the direction of motion of a point (x, y) along the trajectories as t increases. Pictures like that in Fig. 10.1 are often called *phase portraits* of the solutions of the corresponding differential systems.

The equilibrium point at the origin in this case ($q < 0$) is called a *saddle point.*

Because the straight lines in Fig. 10.1 represent, in fact, four actual trajectories (the four rays satisfying the equations $y = 2x$ and $y = -x$, with the origin deleted), these lines could have been determined initially by substituting $y = mx$ in the differential equation

$$\frac{dy}{dx} = \frac{-2x}{x - y}.$$

Such a substitution leads to the quadratic equation

$$m^2 - m - 2 = 0,$$

the roots of which are $m = 2, -1$.

The notion of the *stability* of the origin—that is, of the nature of motions along the trajectories in the neighborhood of the origin—will be discussed in Chapter 11. For the present, we shall say that the origin is *asymptotically stable* when, as in Fig. 10.2, all motions along trajectories approach the origin, as $t \to \infty$. When, however, in every neighborhood of the origin there is at least one point with the property that the trajectory through that point does not remain bounded as $t \to \infty$, the origin will be said to be *unstable*. The origin in Fig. 10.1 is unstable. Indeed, the origin is unstable whenever it is a saddle point.

These definitions apply to linear systems with constant coefficients having the origin as an isolated equilibrium point, but they are consistent with the more general definitions that will be given in Chapter 11.

Case 2: λ_1, λ_2 *real, unequal, and having the same sign.* In this case, $q > 0$, $\Delta > 0$, and the characteristic roots are either both positive or both negative. As in the treatment of Case 1, we shall analyze a typical case. To that end, consider the system

(1.13)
$$\frac{dx}{dt} = -\frac{7}{4}x + \frac{1}{4}y,$$

$$\frac{dy}{dt} = \frac{3}{4}x - \frac{5}{4}y.$$

The characteristic equation is

$$\lambda^2 + 3\lambda + 2 = 0,$$

and the characteristic roots are $\lambda_1 = -1$, $\lambda_2 = -2$. The general solution of (1.13) can be written

(1.14)
$$x = c_1 e^{-t} + c_2 e^{-2t},$$

$$y = 3c_1 e^{-t} - c_2 e^{-2t}.$$

Again, if $c_1 = c_2 = 0$, we obtain the null solution at the origin. If $c_1 = 0, c_2 \neq 0$, as t varies from t_0 to $+\infty$, the point (x, y) moves toward the origin along one of the two rays that form the line $y = -x$ with the origin deleted. Similarly, if $c_1 \neq 0, c_2 = 0$, as t varies from t_0 to $+\infty$, the point (x, y) of a solution moves toward the origin along one of the rays defined by the line $y = 3x$ with the origin deleted.

If the parameter t is eliminated from equation (1.14) or if equation (1.9) is solved, the family of trajectories is given by the equation

(1.15) $$(x + y)^2 = k(y - 3x).$$

These curves are seen, from the form of (1.15), to be tangent to the line $y = 3x$ at the origin (they are parabolas in this particular example, with the lines $2x + 2y + k = 0$ as their axes of symmetry). The equilibrium point at the origin in this case is called a *node*. In this example the node is asymptotically stable. This is clear from equations (1.14).

Again, we observe that there are two straight lines that are trajectories that have equations of the form $y = mx$. These lines could have been determined by substituting $y = mx$ in the differential equation

$$\frac{dy}{dx} = \frac{3x - 5y}{-7x + y}.$$

Such a substitution leads at once to the quadratic equation

$$m^2 - 2m - 3 = 0,$$

the roots of which are $m = 3, -1$.

Further, we observe from (1.14) that

$$\frac{dy}{dx} = \frac{-3c_1e^{-t} + 2c_2e^{-2t}}{-c_1e^{-t} - 2c_2e^{-2t}} = \frac{-3c_1 + 2c_2e^{-t}}{-c_1 - 2c_2e^{-t}}.$$

It follows that, as $t \to +\infty$, the slope of each trajectory approaches 3; that is, all trajectories are tangent to the line $y = 3x$ at the origin, as was noted earlier. More carefully stated, as $t \to +\infty$, the motion along each trajectory is toward the origin, and the slope of each trajectory approaches the slope of the line $y = 3x$.

Further, because

$$\frac{dy}{dx} = \frac{-3c_1e^t + 2c_2}{-c_1e^t - 2c_2}$$

along trajectories, we note that as $t \to -\infty$, $y' \to -1$; that is, as $t \to -\infty$, the slope of each trajectory ($y \ne 3x$) approaches -1.

If the characteristic roots λ_1 and λ_2 had both been positive, the picture would have been similar, but the arrows on the trajectories in Fig. 10.2 would have been reversed, since the movement of a point (x, y) on each trajectory would have been away from the origin, as t increased, and the node would have been unstable.

Case 3: λ_1, λ_2 *conjugate imaginaries* $\alpha \pm i\beta$ ($\beta \ne 0$). In this case, $\Delta < 0$, and if $\alpha \ne 0$, the equilibrium point at the origin is called a *focus*.

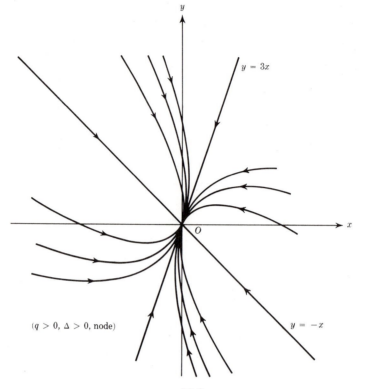

FIG. 10.2

Consider the example

$$\frac{dx}{dt} = x - y,$$

(1.16)

$$\frac{dy}{dt} = x + y.$$

The characteristic roots are $\lambda_1 = 1 + i$, $\lambda_2 = 1 - i$, and the general solution may be written in the form

$$x = e^t(c_1 \cos t + c_2 \sin t)$$
$$y = e^t(c_1 \sin t - c_2 \cos t).$$

The family of curves represented in the phase plane by these equations has the equation

(1.17) $$\sqrt{x^2 + y^2} = ce^{\text{arc tan } y/x},$$

where c is an arbitrary constant. The form of this equation suggests the introduction of polar coordinates ρ and θ. Since in polar coordinates,

$$\rho = \sqrt{x^2 + y^2},$$

$$\tan \theta = \frac{y}{x},$$

equation (1.17) becomes

$$\rho = ce^{\theta},$$

which is the equation of a spiral.

The situation is described schematically in Fig. 10.3 (the actual spirals in this example expand extremely rapidly). The motion is away from the equilibrium point because the real part of the characteristic roots is positive, and the focus is unstable. If the real part of the characteristic roots were negative, the arrow would be reversed in Fig. 10.3, and the focus would be asymptotically stable.

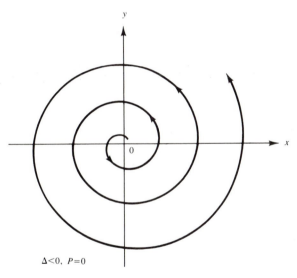

$\Delta < 0, \ P = 0$

FIG. 10.3

When the roots of the characteristic equations are real and unequal, the general case is strictly analogous to the first two cases given above. When the roots are complex conjugates, the general case requires additional analysis.

Consider the system

(1.18)
$$\dot{x} = ax + by,$$
$$\dot{y} = cx + dy \qquad (b \neq 0).$$

Let the characteristic equation

$$\lambda^2 - (a + d)\lambda + (ad - bc) = 0$$

have the roots $\lambda = \alpha \pm i\beta$ $(\beta > 0)$. Then $\alpha^2 + \beta^2 = ad - bc > 0$, and $a + d = 2\alpha$.

To study the phase portrait of the system (1.18) it is convenient to introduce a new coordinate system by means of the substitution

(1.19)

$$x = \frac{1}{\beta} X,$$

$$y = \frac{\alpha - a}{b\beta} X - \frac{1}{b} Y.$$

Then

$$\dot{x} = \frac{1}{\beta} \dot{X} = ax + by = \frac{\alpha}{\beta} X - Y,$$

$$\dot{y} = \frac{\alpha - a}{b\beta} \dot{X} - \frac{1}{b} \dot{Y} = cx + dy = \left[\frac{c}{\beta} + \frac{d(\alpha - a)}{b\beta} \right] X - \frac{d}{b} Y.$$

Solving these equations for $\dot{X}$, $\dot{Y}$ in terms of X, Y we obtain

$$\dot{X} = \alpha X - \beta Y,$$
$$\dot{Y} = \beta X + \alpha Y.$$

We now introduce polar coordinates. Setting

(1.20) $$X = \rho \cos \theta, \qquad Y = \rho \sin \theta$$

we have

(1.21) $$\dot{\rho} \cos \theta - \rho \dot{\theta} \sin \theta = \rho(\alpha \cos \theta - \beta \sin \theta),$$
$$\dot{\rho} \sin \theta + \rho \dot{\theta} \cos \theta = \rho(\beta \cos \theta + \alpha \sin \theta).$$

These equations are readily solved for $\dot{\rho}$, $\dot{\theta}$ yielding

(1.22) $$\dot{\rho} = \alpha\rho,$$
$$\dot{\theta} = \beta. \qquad (\beta > 0).$$

Thus,

$$\frac{d\rho}{d\theta} = \frac{\alpha}{\beta} \rho,$$

and the trajectories in the phase plane are the spirals

(1.23) $$\rho = ke^{(\alpha/\beta)\theta}.$$

Motion along trajectories will be asymptotically stable when $\alpha < 0$, and unstable (spiral expanding) when $\alpha > 0$. When $\alpha = 0$, then $\dot{\rho} = 0$, and the trajectories are the closed curves, $\rho =$ constant; that is, the ellipses

$$(1.24) \qquad (ax + by)^2 + \beta^2 x^2 = k^2,$$

where $k\ (>0)$ is constant. This follows at once from the fact that

$$\rho^2 = X^2 + Y^2,$$

and

$$X = \beta x, \qquad Y = (\alpha - a)x - by.$$

Accordingly, the equation $\rho = k$ is equivalent to equation (1.24).

In the example

$$(1.25) \qquad \begin{aligned} \dot{x} &= -y, \\ \dot{y} &= x. \end{aligned}$$

$a = 0,\ b = -1,\ c = 1,\ d = 0,\ \lambda = \pm i,\ \alpha = 0,\ \beta = 1,\ \rho$ is the usual polar distance $\sqrt{x^2 + y^2}$, and θ is the usual polar angle. The trajectories (1.23) become, of course, the circles

$$x^2 + y^2 = k^2.$$

This system (1.25) is representative of the interesting case when the characteristic roots are pure imaginaries $\pm i\beta\ (\beta > 0)$. Then $\alpha = (a + d)/2 = 0$, and the trajectories are ellipses having the equation

$$(ax + by)^2 + (ad - bc)x^2 = k^2 \qquad (d = -a,\ ad - bc > 0)$$

in the phase plane. Typical is the system

$$\begin{aligned} \dot{x} &= 4x + 5y, \\ \dot{y} &= -5x - 4y. \end{aligned}$$

The characteristic equation is

$$\begin{vmatrix} 4 - \lambda & 5 \\ -5 & -4 - \lambda \end{vmatrix} = \lambda^2 + 9 = 0,$$

and the characteristic roots are $\lambda_1 = 3i$, $\lambda_2 = -3i$. (Note that if this system is regarded as numbers from a physical problem, the slightest change may cause $a + d \neq 0$, and the characteristic roots will no longer be pure imaginaries.)

In this case, the general solution may be written in the form

$$x = c_1(5 \cos 3t) + c_2(5 \sin 3t),$$

$$y = c_1(-4 \cos 3t - 3 \sin 3t) + c_2(3 \cos 3t - 4 \sin 3t).$$

The trajectories in the phase plane have the equation

$$5x^2 + 8xy + 5y^2 = c^2,$$

which is that of a family of concentric ellipses having the origin as center. The equilibrium point at the origin, when $\Delta < 0$, $p = 0$, is called a *center* (see Fig. 10.4). Centers are said to be *stable* for systems of the form (1.1). They are not asymptotically stable.

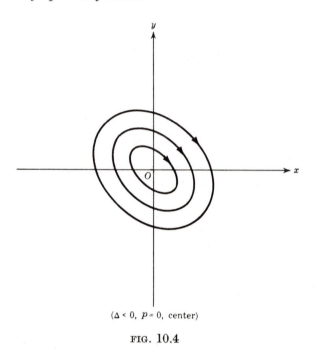

$(\Delta < 0, \ p = 0, \ \text{center})$

FIG. 10.4

When sketching trajectories of systems

$$\dot{x} = ax + by,$$
$$\dot{y} = cx + dy,$$

it is frequently useful to observe that along a ray $y = hx$,

$$\frac{dy}{dx} = \frac{c + dh}{a + bh};$$

that is, all trajectories cross a given ray at the same angle, and in the same direction, since

$$\dot{x} = x(a + dh),$$
$$\dot{y} = x(c + dh).$$

For example, in sketching a typical trajectory for the system (1.16) one observes that the trajectory that crosses the x-axis at $(1, 0)$ does so with $\dot{x} = 1$, $\dot{y} = 1$ at that point. Accordingly, all trajectories have slope 1 where they cross the positive x-axis, and motion at those points is upward and to the right. This observation coupled with the fact that $\alpha > 0$ for (1.16) indicates that each trajectory is expanding as it winds about the origin in a counterclockwise direction.

The discussion of the case $\Delta < 0$ is complete. In addition we have analyzed the cases $\Delta > 0$, $q \neq 0$. There remain

1. $(\Delta > 0)$, $q = 0$, $p \neq 0$;

2. $\Delta = 0$, $p \neq 0$;

3. $(\Delta = 0)$, $q = 0$, $p = 0$.

These cases are less important (in the linear case), and we shall not discuss them.

2 More general systems

Consider the autonomous system

$$(1.26) \qquad \begin{aligned} \dot{x} &= f(x, y), \\ \dot{y} &= g(x, y), \end{aligned}$$

where $f(x, y)$ and $g(x, y)$ are of class C' in an open domain D containing the origin, and $f(0, 0) = g(0, 0) = 0$. If, at the origin

$$\frac{\partial f}{\partial x} = a, \qquad \frac{\partial f}{\partial y} = b,$$

$$\frac{\partial g}{\partial x} = c, \qquad \frac{\partial g}{\partial y} = d,$$

equations (1.26) can be written in the form

$$(1.27) \qquad \begin{aligned} \dot{x} &= ax + by + F(x, y), \\ \dot{y} &= cx + dy + G(x, y). \end{aligned}$$

The linear equations

$$(1.28) \qquad \begin{aligned} \dot{x} &= ax + by, \\ \dot{y} &= cx + dy \end{aligned}$$

are called the *equations of variation* associated with (1.26) and (1.27). It is natural to ask under what conditions the behavior of the trajectories of (1.28) near the origin is representative of the behavior of the trajectories of (1.27) near the origin. It is of particular importance to ascertain conditions under which the stability of the origin in (1.28) implies stability of the origin for (1.27). We shall examine questions of this kind in the next chapter.

Exercises

Discuss the systems in Exercises 1–12. Sketch typical trajectories and determine the stability or instability of the origin.

1. $\dfrac{dx}{dt} = x + y,$

$\dfrac{dy}{dt} = 3x - y.$

2. $\dfrac{dx}{dt} = x - y,$

$\dfrac{dy}{dt} = 2x + 4y.$

3. $\dfrac{dx}{dt} = -x - y,$

$\dfrac{dy}{dt} = x - y.$

4. $\dfrac{dx}{dt} = 3x + 5y,$

$\dfrac{dy}{dt} = -5x - 3y.$

5. $\dfrac{dx}{dt} = 2x + y,$

$\dfrac{dy}{dt} = -5x - 4y.$

6. $\dfrac{dx}{dt} = x - 3y,$

$\dfrac{dy}{dt} = 4x - 6y.$

7. $\dfrac{dx}{dt} = 3x - 5y,$

$\dfrac{dy}{dt} = 2x + y.$

8. $\dfrac{dx}{dt} = 3x - 13y,$

$\dfrac{dy}{dt} = x - 3y.$

9. $\dot{x} = -2x + y,$
$\dot{y} = 2x - 3y.$

10. $\dot{x} = y,$
$\dot{y} = 2x + y.$

11. $\dot{x} = 2x + 2y,$
$\dot{y} = x + 3y.$

12. $\dot{x} = 2x + 3y,$
$\dot{y} = 3x + 2y.$

13. The differential equation

$$\frac{d^2x}{dt^2} + \mu(x^2 - 1)\frac{dx}{dt} + x = 0 \qquad (\mu \text{ a positive constant})$$

is known as van der Pol's equation. It is of importance in vacuum tube theory. The substitution $v = dx/dt$ leads to the pair of equations

(1)

$$\frac{dx}{dt} = v,$$

$$\frac{dv}{dt} = -x + \mu(1 - x^2)v.$$

The equations of variation of the system (1) are the associated linear equations

(2)

$$\frac{dx}{dt} = v,$$

$$\frac{dv}{dt} = -x + \mu v.$$

Discuss the system (2) and draw some typical trajectories in the xv-plane (the phase plane). Consider two cases: $\mu < 2$, $\mu > 2$. (The case $\mu = 2$ is more complicated and may be omitted for present purposes.)

14. Transform the following pairs of equations by means of (1.19) and (1.20) and obtain equations corresponding to (1.22) and to (1.23).

(a) $\dot{x} = 2x + y,$
$\dot{y} = -x + 2y;$

(b) $\dot{x} = -2x + y,$
$\dot{y} = -x - 2y;$

(c) $\dot{x} = -2x + y,$
$\dot{y} = -5x + 2y;$

(d) $\dot{x} = -x - y,$
$\dot{y} = 10x - 3y;$

(e) $\dot{x} = y,$
$\dot{y} = -2x - 2y.$

15.* Prove that if $\lambda \neq \lambda_2$, the solutions (1.7) are linearly independent (assume $b \neq 0$).

Answers

1. $\lambda = 2, -2;$ saddle point; unstable.

2. $\lambda = 2, 3;$ node; unstable.

3. $\lambda = -1 \pm i;$ focus; asymptotically stable.

4. $\lambda = 4i, -4i;$ center; stable.

6. $\lambda = -2, -3;$ node; stable.

8. $\lambda = \pm 2i;$ center; stable.

10. $\lambda = 2, -1;$ saddle point; unstable.

12. $\lambda = 5, -1;$ saddle point; unstable.

14. (a) $x = X, \quad y = -Y;$
$\qquad \rho = ke^{2\theta}.$

(c) $x = X, \quad y = 2X - Y;$
$\qquad \rho = $ constant.

(e) $x = X, \quad y = -X - Y;$
$\qquad \rho = ke^{-\theta}.$

11

Stability; Liapunov's direct method[†]

1 Nonlinear differential systems

An enormous number of practical problems in the control of electrical, chemical, biological, mechanical, thermal, and other systems may be reduced to the study of a system of differential equations of the type

$$\frac{dx_1}{dt} = X_1(x_1, x_2, \ldots, x_n, t),$$

(1.1)
$$\frac{dx_2}{dt} = X_2(x_1, x_2, \ldots, x_n, t),$$

$$. \quad . \quad . \quad . \quad . \quad . \quad . \quad . \quad .$$

$$\frac{dx_n}{dt} = X_n(x_1, x_2, \ldots, x_n, t),$$

---

† Also called *Liapunov's second method*.

where the quantities X_i are functions, not necessarily linear, of the $n + 1$ variables $x_1, x_2, \ldots, x_n, t$. If $x_1, x_2, \ldots, x_n$ are regarded as components of an n-vector x and $X_1, X_2, \ldots, X_n$ as components of an n-vector X, equation (1.1) may be written in abbreviated form as

$$(1.2) \qquad \frac{dx}{dt} = X(x, t).$$

Recall that when the variable t does not appear formally in the right-hand member of (1.1), the system (1.1) is said to be an *autonomous* system; otherwise, it is called *nonautonomous*. When (1.2) is autonomous, it can be rewritten as

$$(1.2)' \qquad \frac{dx}{dt} = X(x).$$

A solution

$$x_1 = x_1(t),$$
$$x_2 = x_2(t),$$
$$. \quad . \quad .$$
$$x_n = x_n(t)$$

of $(1.2)'$ may be regarded either as a curve in the space of the $n + 1$ variables $x_1, x_2, \ldots, x_n, t$ or as a curve in the space of n variables $x_1, x_2, \ldots, x_n$ with t regarded as a parameter. In the latter case (which will be our principal concern), the curve is called a *trajectory* in the *phase space*—the space of the n variables $x_1, x_2, \ldots, x_n$.

For example, the autonomous system

$$(1.3) \qquad \frac{dx_1}{dt} = x_2,$$
$$\frac{dx_2}{dt} = -x_1$$

has the solution

$$(1.4) \qquad x_1 = \sin t,$$
$$x_2 = \cos t.$$

The solution (1.4) will be recognized as a helix in (x_1, x_2, t)-space, while the corresponding trajectory in the phase space (i.e. the $x_1 x_2$-plane) is the circle $x_1^2 + x_2^2 = 1$. In the latter case t is regarded as a parameter.

Any point $(a_1, a_2, \ldots, a_n)$ that provides a solution of the system of equations

$$X_1(x_1, x_2, \ldots, x_n) = 0,$$
$$. \quad . \quad . \quad . \quad . \quad .$$
$$X_n(x_1, x_2, \ldots, x_n) = 0$$

is called an *equilibrium point* of the system $(1.2)'$. It is important to study the behavior of trajectories in the neighborhood of an equilibrium point and for this purpose it is usually convenient to assume that the equilibrium point is at the origin. The translation

$$x_i = a_i + z_i \qquad (i = 1, 2, \ldots, n)$$

transforms the system

(1.5) $$\frac{dx}{dt} = X(x)$$

into the equivalent system

(1.6) $$\frac{dz}{dt} = Z(z),$$

where

$$Z(z) = X(a + z),$$

and an equilibrium point $(a_1, a_2, \ldots, a_n)$ of the system (1.5) becomes the equilibrium point $(0, 0, \ldots, 0)$ of the system (1.6).

Exercises

1. Find the solution of the linear system

$$\frac{dx_1}{dt} = x_2,$$

$$\frac{dx_2}{dt} = -4x_1,$$

with the property that $x_1(0) = 0$, $x_2(0) = 1$. Interpret the solution in (x_1, x_2, t)-space and find the corresponding trajectory in the phase plane.

2. Find the solution of the linear system

$$\frac{dx_1}{dt} = x_1,$$

$$\frac{dx_2}{dt} = 2x_2$$

with the property that $x_1(0) = 1$, $x_2(0) = 1$. Interpret the solution in (x_1, x_2, t)-space and find the corresponding trajectory in the phase plane.

3. Show that the equilibrium points of the nonlinear system

$$\frac{dx_1}{dt} = 8x_1 - x_2^2,$$

$$\frac{dx_2}{dt} = x_2 - x_1^2$$

are the points $(0, 0)$ and $(2, 4)$. Transform the system to an equivalent system in which the equilibrium point $(2, 4)$ has been translated to the origin.

4. Show that the equilibrium points of the nonlinear system

$$\frac{dx_1}{dt} = 5 - x_1^2 - x_2^2,$$

$$\frac{dx_2}{dt} = x_2 - 2x_1$$

are the points $(1, 2)$ and $(-1, -2)$. Transform the system to an equivalent system in which the equilibrium point $(-1, -2)$ has been translated to the origin.

5. Find the equilibrium points of the nonlinear system

$$\frac{dx_1}{dt} = x_1 + x_2 - 5,$$

$$\frac{dx_2}{dt} = x_1 x_2 - 6.$$

6. Find the equilibrium points of the nonlinear system

$$\frac{dx_1}{dt} = 293 - x_1^2 - x_2^2,$$

$$\frac{dx_2}{dt} = x_1 x_2 - 34.$$

7. Find the equilibrium points of the nonlinear system

$$\frac{dx_1}{dt} = x_1(x_2 + x_3) - 12,$$

$$\frac{dx_2}{dt} = 6 - x_2(x_1 + x_3),$$

$$\frac{dx_3}{dt} = x_3(x_1 + x_2) - 10.$$

Answers

2. $x_1 = e^t$, $x_2 = e^{2t}$; a curve lying on the parabolic cylinder $x_2 = x_1^2$; the parabola $x_2 = x_1^2$.

3. The transformation $x_1 = 2 + z_1$, $x_2 = 4 + z_2$ leads to the equivalent system

$$\frac{dz_1}{dt} = 8z_1 - 8z_2 - z_2^2,$$

$$\frac{dz_2}{dt} = z_2 - 4z_1 - z_1^2.$$

7. $(4, 1, 2)$, $(-4, -1, -2)$.

2   Stability and instability

In the remainder of this chapter we shall deal with autonomous systems of the form

$$\frac{dx_1}{dt} = X_1(x_1, x_2, \ldots, x_n),$$

$$\frac{dx_2}{dt} = X_2(x_1, x_2, \ldots, x_n),$$

(2.1)

$$\cdot \quad \cdot \quad \cdot \quad \cdot \quad \cdot \quad \cdot \quad \cdot$$

$$\frac{dx_n}{dt} = X_n(x_1, x_2, \ldots, x_n),$$

where all partial derivatives $\partial X_i/\partial x_j$ exist and are continuous in an open n-dimensional region D containing the origin. This region will often be all of n-space, which is the phase space of the n variables $x_1, x_2, \ldots, x_n$. Convenient shorthand for (2.1) is

(2.1)′
$$\frac{dx}{dt} = X(x).$$

We shall assume further that $X(0) = 0$ and that the origin is an *isolated equilibrium point* of the system (2.1)—that is, there exists a sphere with center at the origin such that no other point for which $X(x)$ is zero lies inside the sphere. The fundamental existence theorem will insure that through each point of D that is not an equilibrium point there exists a unique trajectory of the system (2.1). We shall suppose this trajectory exists for all values of t greater than or equal to some value t_0.

We shall be concerned with motions along trajectories in the phase space. Such studies were initiated in Chapter 10 by considering motions along trajectories in the xy-plane for linear systems such as

$$\frac{dx}{dt} = y,$$

(2.2)

$$\frac{dy}{dt} = -2x - 3y.$$

In the present chapter, we shall be particularly concerned with the stability or instability of such motions† in a neighborhood of the equilibrium point at the origin. We begin with some definitions.

---

† The reader is referred to the book by J. P. LaSalle and S. Lefschetz, *Stability by Liapunov's Direct Method, with Applications*, Academic Press, New York (1961), for more information on this and other matters pertaining to the material of this chapter. See also the author's *An Introduction to the Theory of Ordinary Differential Equations*, Wadsworth Publishing Company, Belmont, California (1976).

The equilibrium point at the origin of the system (2.1) is said to be *stable* if corresponding to each positive number R there exists a positive number r $(r \le R)$ such that a trajectory which at time $t = t_0$ is at a point inside the sphere

$$x_1^2 + x_2^2 + \cdots + x_n^2 = r^2$$

remains inside the sphere

$$x_1^2 + x_2^2 + \cdots + x_n^2 = R^2$$

for all $t > t_0$. The situation is shown schematically in Fig. 11.1.

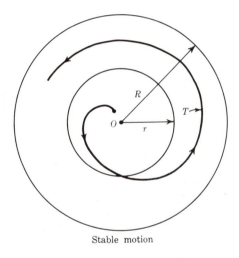

Stable motion

FIG. 11.1

The equilibrium point at the origin is said to be *asymptotically stable* in an open domain D containing the origin if it is stable and if every trajectory that starts in D at time $t = t_0$ tends to the origin as $t \to +\infty$. The domain D is then said to be a *domain of attraction* of the equilibrium point at the origin. (See Fig. 11.2.)

The equilibrium point at the origin is said to be *unstable* in a region D containing the origin when, for any positive number r, however small, there is some point P inside the sphere $x_1^2 + x_2^2 + \cdots + x_n^2 = r^2$ such that the trajectory that is at P at time t_0 will reach the boundary of D at time t_1, where t_1 is some number **greater than** t_0.

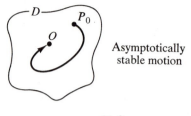

Asymptotically
stable motion

FIG. 11.2

When a trajectory is a closed curve, the motion along that trajectory is said to be *periodic*. We observe that the equilibrium point at the origin is stable in Fig. 10.4, unstable in Figs. 10.1 and 10.3, and asymptotically stable in Fig. 10.2. The trajectories in Fig. 10.4 are also examples of periodic motions.

3   Characteristic roots

In this section we shall recall briefly some of the ideas developed in Chapter 10.

To solve the linear system with constant coefficients

(3.1)
$$\dot{x}_1 = a_{11}x_1 + a_{12}x_2 + \cdots + a_{1n}x_n,$$
$$\dot{x}_2 = a_{21}x_1 + a_{22}x_2 + \cdots + a_{2n}x_n,$$
$$\cdot \quad \cdot \quad \cdot \quad \cdot \quad \cdot \quad \cdot \quad \cdot \quad \cdot \quad \cdot \quad \cdot \quad \cdot$$
$$\dot{x}_n = a_{n1}x_1 + a_{n2}x_2 + \cdots + a_{nn}x_n,$$

recall that we try for a solution of the form

$$x_1 = A_1 e^{\lambda t},$$
$$x_2 = A_2 e^{\lambda t},$$
$$\cdot \quad \cdot \quad \cdot \quad \cdot$$
$$x_n = A_n e^{\lambda t}.$$

This leads to the *characteristic equation*

(3.1)′
$$\begin{vmatrix} a_{11} - \lambda & a_{12} & \cdots & a_{1n} \\ a_{21} & a_{22} - \lambda & \cdots & a_{2n} \\ \cdot & \cdot & \cdot & \cdot \\ a_{n1} & a_{n2} & \cdots & a_{nn} - \lambda \end{vmatrix} = 0.$$

This is a polynomial equation of degree n in λ with real coefficients. The variety of possible roots of such an equation is well known, and the types of terms that appear in the general solution of (3.1) are of the form

$$(3.2) \qquad\qquad e^{\lambda t} P(t),$$

where λ is a characteristic root (possibly imaginary) and $P(t)$ is a polynomial in t. If some root λ has a positive real part (for example, when $\lambda = 2$, or when $\lambda = 3 + 4i$), some solutions, at least, will not remain bounded as t approaches ∞, and the origin will be unstable.† If, however, every characteristic root has a negative real part, every term (3.2) in the solution will tend to 0, as t approaches ∞, and the origin will be asymptotically stable.

Consider now the nonlinear system

$$(3.3) \qquad \begin{aligned} \dot{x} &= y, \\ \dot{y} &= -2x - 3y + x^3. \end{aligned}$$

The linear terms in (3.3) yield the so-called equations of variation

$$(3.4) \qquad \begin{aligned} \dot{x} &= y, \\ \dot{y} &= -2x - 3y. \end{aligned}$$

The origin is asymptotically stable for the system (3.4), since the characteristic roots $\lambda = -1, -2$ are negative. Inasmuch as the nonlinear term in (3.3) is small near the origin, we would expect that the system (3.4) would determine the stability of the origin in (3.3). It does, and the origin in (3.3) is asymptotically stable. Liapunov theory, introduced in the following section, provides a simple and elegant way of dealing with systems like (3.3). It is effective also in dealing with linear systems having constant coefficients, largely because the task of determining the characteristic roots—or even the signs of their real parts—is often formidable when the characteristic equation is other than quadratic.

Consider, also, the system

$$(3.5) \qquad \begin{aligned} \dot{x} &= y, \\ \dot{y} &= -3y + kx^3, \end{aligned}$$

where k is a constant $\neq 0$. In this case, the equations of variation are

$$(3.6) \qquad \begin{aligned} \dot{x} &= y, \\ \dot{y} &= -3y, \end{aligned}$$

---

† Recall Euler's formula, Chapter 3, equation (2.7).

and while the origin is an isolated equilibrium point for (3.5), it is not for (3.6), for every point of the x-axis is an equilibrium point for (3.6). In the latter case, the characteristic equation is, of course,

$$\begin{vmatrix} -\lambda & 1 \\ 0 & -3-\lambda \end{vmatrix} = 0,$$

and the characteristic roots are $\lambda = 0,\ -3$. The characteristic roots do not give us enough information to determine the stability of the origin for the system (3.5), in which the nonlinearity is the decisive factor. Liapunov theory will tell us rather quickly that when $k < 0$, the origin is asymptotically stable and, equally quickly, that the origin is unstable when $k > 0$.

4   Positive definite functions

We are ready to formulate the criteria that determine whether or not an equilibrium point is stable or unstable. To that end, let

$$V(x) = V(x_1, x_2, \ldots, x_n)$$

be a function of class C' in an open region H containing the origin. Suppose $V(0) = 0$ and that V is positive at all other points of H. Then V has a minimum at the origin, and V is said to be *positive definite* in H. Clearly, the origin is a *critical point* of V—that is, a point at which all the partial derivatives

$$\frac{\partial V}{\partial x_1}, \frac{\partial V}{\partial x_2}, \ldots, \frac{\partial V}{\partial x_n}$$

vanish. The origin will be said to be an *isolated* critical point if there is a sphere with center at the origin such that the origin is the only critical point of V inside the sphere.

Before looking at the geometry in n dimensions let us consider the case $n = 2$. When $V(x, y)$ is positive definite and has the origin as an isolated critical point, it is intuitively evident that the curves

$$V(x, y) = k,$$

for k positive and sufficiently small, define ovals (closed curves) containing the origin in their interior. Further, when $k_1 < k_2$, the oval defined by $V(x, y) = k_1$ lies entirely within the oval defined by $V(x, y) = k_2$ (see Fig. 11.3). This can be shown to be true, in general, but the proof is not elementary.

We state without proof the *nesting property* of such functions in n-space.

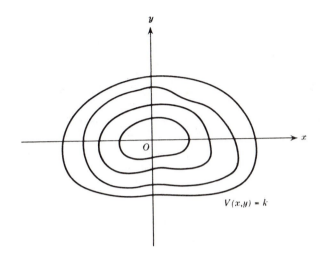

FIG. 11.3

When

$$V(x) = V(x_1, x_2, \ldots, x_n)$$

is positive definite in H and the minimum at the origin is an isolated critical point of $V(x)$, there exists a positive number k_0 with the property that the equation

$$V(x) = k$$

defines a closed surface containing the origin in its interior, if $0 < k < k_0$. Further, the closed surface $V(x) = k_2$ lies wholly within the surface $V(x) = k_1$, if $0 < k_2 < k_1 < k_0$. Finally, at each point inside such a closed surface defined by $V(x) = k$, the value of V is positive but less than k.

Example. Consider the function

$$V(x, y) = 3x^2 + y^2 - x^3.$$

It is positive definite neighboring the origin, and its only critical points are at $(0, 0)$ and $(2, 0)$. Thus, the critical point at the origin is isolated. It can be shown that for this function V, the number $k_0 = 4$. Consider the curve

(4.1) $$3x^2 + y^2 - x^3 = 2.$$

Its graph is given in Fig. 11.4.

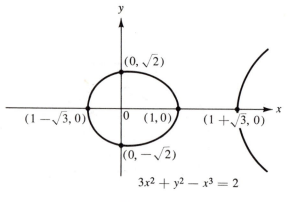

FIG. 11.4

Note that the total curve (4.1) is not an oval, but the equation defines an oval near the origin. The equation

$$3x^2 + y^2 - x^3 = 1,$$

for example, will then define an oval lying wholly within the oval defined by (4.1) and containing the origin in its interior.

Henceforth, we shall say that a function $V(x)$ is *positive definite* in H, if the following conditions are satisfied:

(a) $V(x)$ is of class C' in H;
(b) $V(0) = 0$;
(c) $V(x)$ is positive, except at the origin;
(d) the origin is an isolated critical point of $V(x)$.

Exercises

In each of the following exercises plot the four curves whose equations are given using the same coordinate axes. In each case the function of x, y involved is positive definite near the origin and has the origin as an isolated minimum point.

1. $x^2 + 2y^2 = k$ $(k = 1, 2, 4, 6)$.

2. $3x^2 + y^2 - x^3 = k$ $(k = 1, 2, 4, 6)$.

3. $\dfrac{3x^2}{1 + x^2} + y^2 = k$ $(k = 1, 2, 3, 4)$.

5   The Liapunov function

To define a Liapunov function we require the following important concept. Suppose we have given an autonomous system (2.1)

(5.1)
$$\dot{x}_1 = X_1(x_1, x_2, \ldots, x_n),$$
$$\dot{x}_2 = X_2(x_1, x_2, \ldots, x_n),$$
$$\cdot \quad \cdot \quad \cdot \quad \cdot \quad \cdot \quad \cdot \quad \cdot$$
$$\dot{x}_n = X_n(x_1, x_2, \ldots, x_n),$$

and a function $V(x) = V(x_1, x_2, \ldots, x_n)$ of class C'. The *derivative* $\dot{V}$ *of* V *along trajectories* of (5.1) is defined by the equation

$$\dot{V} = \frac{dV}{dt} = \frac{\partial V}{\partial x_1} X_1(x) + \frac{\partial V}{\partial x_2} X_2(x) + \cdots + \frac{\partial V}{\partial x_n} X_n(x).$$

The classical definition of a Liapunov function is then this: *if* $V(x)$ *is positive definite in H, and if* $\dot{V} \leq 0$ *throughout H, then* $V(x)$ *is a Liapunov function for the equilibrium point at the origin of the system (5.1)—or, when there is no ambiguity, simply a Liapunov function for the system (5.1).*

We may now prove the following fundamental result.

Theorem 5.1.   (LIAPUNOV) *If there exists a Liapunov function for (5.1), the equilibrium point at the origin is stable.*

To prove the theorem let S_1 be a sphere of radius R_1 with center at the origin lying in H. Next, choose a positive number k such that the closed surface S defined by $V(x) = k$ lies inside S_1, and then let S_0 be a sphere of radius R_0 ($R_0 < R$) with center at the origin that lies inside S. The situation is indicated schematically in Fig. 11.5.

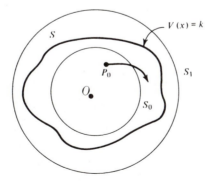

FIG. 11.5

Suppose, then, a trajectory is at P_0 inside S_0 at time t_0. Since P_0 lies inside S, it follows that $V(P_0) < k$. Further, since $\dot{V} \le 0$, V never increases along a trajectory as t increases. Accordingly, the trajectory can never reach the surface $V(x) = k$, and consequently cannot reach S_1.

Liapunov functions may be regarded as extensions of the idea of energy function in mechanics. But to apply Liapunov's idea, any function $V(x, y)$ satisfying the hypotheses of the theorem will suffice. In simpler cases we may often guess usable Liapunov functions; in more complicated cases determining a Liapunov function for a given system may be very difficult indeed.

Example.   Consider the system (2.2)

$$\frac{dx}{dt} = y,$$

(5.2)

$$\frac{dy}{dt} = -2x - 3y.$$

If we set

$$V(x, y) = 2x^2 + y^2,$$

then

$$\dot{V} = 2y(-2x - 3y) + 4x(y)$$
$$= -6y^2.$$

The function $V(x, y)$ is, accordingly, a Liapunov function for (5.2), and the origin is stable.

As a second example consider the system

$$\frac{dx}{dt} = y.$$

$$\frac{dy}{dt} = -x.$$

The point $(0, 0)$ is the only equilibrium point of the system. A suitable Liapunov function is $V(x, y) = x^2 + y^2$. For then

$$\dot{V} = 2x(y) + 2y(-x) = 0,$$

and the critical point at the origin is stable. The trajectories are readily seen to be the family of circles $x^2 + y^2 = a^2$.

Theorem 5.2.   (LIAPUNOV) *If in addition to the hypotheses of Theorem 5.1, $-V \in C'$ is positive definite, the equilibrium point at the origin is asymptotically stable.*

Note first that the origin is stable.

In this case, as t increases, V decreases steadily along each trajectory and, hence, approaches some limiting value $k_1 \geq 0$ on any given trajectory. Because V is continuous along the trajectory, it is sufficient to show that $k_1 = 0$.

Suppose, if possible, that $k_1 > 0$. Then the trajectory could not penetrate the closed surface S_1 determined by the equation $V = k_1$. Next, let $\delta > 0$ be taken small enough that the closed surface D determined by the equation $-\dot{V} = \delta$ lies wholly within the surface S_1. Recall that $-\dot{V}$ is less than δ inside D and is greater than δ outside D. Then,

(5.3) $-\dot{V} > \delta$

on the given trajectory for all t sufficiently large, say $\geq t_0$. Integrating the inequality (5.3), we would have

$$\int_{t_0}^{t} \dot{V}\, dt < -\int_{t_0}^{t} \delta\, dt,$$

or

$$V(t) - V(t_0) < -\delta(t - t_0);$$

that is, for t large enough, $V(t)$ would be negative. But this is impossible, and the theorem is established.

The student should convince himself that δ can always be chosen so that D lies wholly within S_1.

As an example of the application of Theorem 5.2 consider the system (see Fig. 10.2)

$$\frac{dx}{dt} = -\tfrac{7}{4}x + \tfrac{1}{4}y,$$

$$\frac{dy}{dt} = \tfrac{3}{4}x - \tfrac{5}{4}y.$$

The origin is the only equilibrium point. A suitable Liapunov function is

$$V(x, y) = 12x^2 + 12xy + 20y^2.$$

The equations $V(x, y) = k$ $(k > 0)$ represent a family of ellipses. We compute

$$\dot{V} = -(33x^2 + 47y^2);$$

accordingly, $\dot{V}$ is negative definite along trajectories, and Theorem 5.2 states that the equilibrium point at the origin is asymptotically stable.

It will be observed that an equally good choice (among others) for V would have been $3x^2 + 6xy + 11y^2$.

We continue with the following theorem on instability.

Theorem 5.3.   (LIAPUNOV) *Let $V(x)$ be of class C'' in a region H containing the origin and suppose that $V(0) = 0$. If $\dot{V}$ (evaluated along trajectories) is positive definite in H, and if in every neighborhood of the origin there is a point where $V > 0$, the equilibrium point at the origin is unstable.*

The proof of the theorem can now be made as follows (see Fig. 11.6). Let $S(R)$ and $S(r)$ be any two spheres in H, centered at the origin, of radius R and r, respectively, with $0 < r < R$. Pick a point P_0 inside $S(r)$ with the property that $V(P_0) > 0$ and consider the trajectory g commencing at P_0 at time $t = t_0$. Since $\dot{V} > 0$ in and on $S(R)$, except at the origin, V increases steadily on g as t increases. Because of the continuity of V there is a ball B, centered at the origin, throughout which $|V| \leq V(P_0)/2$. It is clear that g cannot reach any point of B. Next, take $\delta > 0$ and so small that the closed surface determined by $\dot{V} = \delta$ lies wholly within B. Applying the nesting property of the function $\dot{V}$ we observe that on g

$$\dot{V} > \delta \qquad (t \geq t_0)$$

and, hence,

$$\int_{t_0}^t \dot{V}\, dt > \int_{t_0}^t \delta\, dt,$$

or, along g,

$$V(t) - V(t_0) > \delta(t - t_0) \qquad (t \geq t_0).$$

Because of the continuity of V in the closed ball bounded by $S(R)$, V is bounded there. On the other hand, by the last inequality, V cannot remain bounded. Thus, g passes outside the sphere $S(R)$, and the equilibrium point at the origin is unstable.

To illustrate the theorem consider the system (see Fig. 10.1)

$$\frac{dx}{dt} = x - y,$$

$$\frac{dy}{dt} = -2x.$$

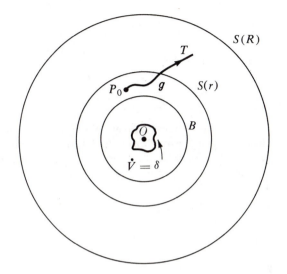

FIG. 11.6

The function $V(x, y) = x^2 - 2xy - y^2$ satisfies the hypotheses of the theorem, for $V > 0$ at every point on the line $2x + y = 0$, except the origin, and along each trajectory

$$\dot{V} = 2x\dot{x} - 2x\dot{y} - 2y\dot{x} - 2y\dot{y}$$
$$= 6x^2 + 2y^2.$$

Accordingly, the equilibrium point at the origin is unstable.

Exercises

For each of the given systems use the suggested function $V(x, y)$ to test the equilibrium point at the origin for stability.

1. $\dfrac{dx}{dt} = -x - y,$

    $\dfrac{dy}{dt} = x - y;$ $2V = x^2 + y^2.$

2. $\dfrac{dx}{dt} = x + y,$

    $\dfrac{dy}{dt} = 3x - y;$ $2V = x^2 + 2xy - y^2.$

3. $\dfrac{dx}{dt} = 3x + 5y,$

$\dfrac{dy}{dt} = -5x - 3y;$ $2V = 5x^2 + 6xy + 5y^2.$

4. $\dot{x} = x - y,$
$\dot{y} = 2x + 4y;$ $2V = 23x^2 + 8xy + 3y^2.$

In the next four exercises find a function $V(x, y)$ that will determine the stability or instability of the equilibrium point at the origin. [*Hint.* Determine constants a, b, c in $V(x, y) = ax^2 + 2bxy + cy^2$. Recall that V is definite when $b^2 - ac < 0$, and is positive definite when $a > 0$ and $b^2 - ac < 0$].

5. $\dot{x} = 3x + 2y,$
$\dot{y} = -7x - 3y.$

6. $\dot{x} = -x - \frac{2}{3}y,$
$\dot{y} = \frac{3}{2}x - y.$

7. $\dfrac{dx}{dt} = -x + y,$

$\dfrac{dy}{dt} = -2x - 4y.$

8. $\dot{x} = x - y,$
$\dot{y} = x + y.$

6 A special case

When the system is linear with constant coefficients—that is, when it has the form

(6.1) $\dot{x} = Ax,$

where

$$A = \begin{bmatrix} a_{11} & a_{12} & \cdots & a_{1n} \\ a_{21} & a_{22} & \cdots & a_{2n} \\ \cdot & \cdot & \cdot & \cdot \\ a_{n1} & a_{n2} & \cdots & a_{nn} \end{bmatrix}, \det A \neq 0.$$

It can be shown† that if the equilibrium point at the origin in (6.1) is asymptotically stable, the matrix equation

(6.2) $A*B + BA = -I$

---

† See, for example, LaSalle and Lefschetz, *loc. cit.*, p. 79.

has a unique solution B, where B is a symmetric matrix. Here, I is the identity matrix

$$\begin{bmatrix} 1 & 0 & 0 & \cdots & 0 \\ 0 & 1 & 0 & \cdots & 0 \\ \cdot & \cdot & \cdot & \cdot & \cdot \\ 0 & 0 & 0 & \cdots & 1 \end{bmatrix},$$

and A^* is the transpose of A. It can be shown further that the resulting quadratic form

$$V = x^* B x$$

is positive definite and is, in fact, a Liapunov function for the system (6.1). Accordingly, if the origin is, in fact, asymptotically stable, this can be determined to be so without solving the characteristic equation (3.1)'.

To illustrate the theory, consider the system

(6.3)
$$\begin{aligned} \dot{x}_1 &= x_1 + x_2, \\ \dot{x}_2 &= -5x_1 - 3x_2. \end{aligned}$$

Here,

$$A = \begin{bmatrix} 1 & 1 \\ -5 & -3 \end{bmatrix},$$

and equation (6.2) becomes

(6.4)
$$\begin{bmatrix} 1 & -5 \\ 1 & -3 \end{bmatrix} \cdot B + B \cdot \begin{bmatrix} 1 & 1 \\ -5 & -3 \end{bmatrix} = -\begin{bmatrix} 1 & 0 \\ 0 & 1 \end{bmatrix}.$$

If we set

$$B = \begin{bmatrix} b_{11} & b_{12} \\ b_{12} & b_{22} \end{bmatrix},$$

equation (6.4) becomes, after carrying out the indicated operations,

$$\begin{bmatrix} 2b_{11} - 10b_{12} & b_{11} - 2b_{12} - 5b_{22} \\ b_{11} - 2b_{12} - 5b_{22} & 2b_{12} - 6b_{22} \end{bmatrix} = \begin{bmatrix} -1 & 0 \\ 0 & -1 \end{bmatrix}.$$

We have, then, the following three equations to determine the numbers b_{ij}:

$$\begin{aligned} 2b_{11} - 10b_{12} &= -1, \\ b_{11} - 2b_{12} - 5b_{22} &= 0, \\ 2b_{12} - 6b_{22} &= -1. \end{aligned}$$

These equations are readily solved, and we find that

$$b_{11} = \tfrac{9}{2}, \qquad b_{12} = 1, \qquad b_{22} = \tfrac{1}{2}.$$

The theory then indicates that

$$V = x^*Bx = \begin{bmatrix} x_1 \\ x_2 \end{bmatrix} \begin{bmatrix} \frac{9}{2} & 1 \\ 1 & \frac{1}{2} \end{bmatrix} [x_1 \quad x_2]$$

$$= \tfrac{9}{2}x_1^2 + 2x_1x_2 + \tfrac{1}{2}x_2^2$$

is a positive definite quadratic form that is a Liapunov function for the system (6.3), if the origin is asymptotically stable. We write

$$2V = 9x_1^2 + 4x_1x_2 + x_2^2,$$

and note that V is, indeed, positive definite. We calculate

$$\dot{V} = 9x_1(x_1 + x_2) + 2x_1(-5x_1 - 3x_2)$$
$$+ 2x_2(x_1 + x_2) + x_2(-5x_1 - 3x_2)$$
$$= -[x_1^2 + x_2^2];$$

consequently, the origin is asymptotically stable.

As a second example, consider the system

$$\dot{x}_1 = x_2,$$
$$\dot{x}_2 = -x_1 + 2x_2.$$

Here,

$$A = \begin{bmatrix} 0 & 1 \\ -1 & 2 \end{bmatrix}.$$

We set

$$B = \begin{bmatrix} b_{11} & b_{12} \\ b_{12} & b_{22} \end{bmatrix},$$

and, after some calculation, we have

$$-2b_{12} \qquad = -1,$$
$$b_{11} + 2b_{12} - b_{22} = 0,$$
$$2b_{12} + 4b_{22} = -1.$$

It follows that

$$b_{11} = -\tfrac{3}{2}, \qquad b_{12} = \tfrac{1}{2}, \qquad b_{22} = -\tfrac{1}{2},$$

and

$$2V = -3x_1^2 + 2x_1x_2 - x_2^2.$$

Inasmuch as V is not positive definite, the origin cannot be asymptotically stable.

Finally, consider the system

$$\dot{x} = x + y,$$
$$\dot{y} = -3x - y.$$

Here,

$$A = \begin{bmatrix} 1 & 1 \\ -3 & -1 \end{bmatrix},$$

and the matrix equation

$$A*B + BA = -I$$

yields the equations

$$
\begin{aligned}
2b_{11} - 6b_{12} & = -1, \\
b_{11} \qquad\quad - 3b_{22} & = \;\;0, \\
2b_{12} - 2b_{22} & = -1.
\end{aligned}
$$

These equations are inconsistent and we conclude that the origin is not asymptotically stable.

The origin is, however, stable, for it is a center. The methods of this section furnish no information in this situation beyond the fact that the origin is not *asymptotically* stable.

When one uses quadratic forms as possible Liapunov functions it is desirable to have a test to determine whether or not the quadratic form is positive definite. Sylvester's theorem from algebra provides an answer. Suppose the quadratic form $V = x*Bx$ has the (symmetric) matrix

$$
B = \begin{bmatrix}
b_{11} & b_{12} & \dots & b_{1n} \\
b_{12} & b_{22} & \dots & b_{2n} \\
\cdot & \cdot & \cdot & \cdot \\
b_{1n} & b_{2n} & \dots & b_{nn}
\end{bmatrix}.
$$

Sylvester's theorem states that *a necessary and sufficient condition that the quadratic form be positive definite is that the n principal determinants of B be positive*; that is, the determinants

$$
|b_{11}|, \quad
\begin{vmatrix} b_{11} & b_{12} \\ b_{12} & b_{22} \end{vmatrix}, \quad
\begin{vmatrix} b_{11} & b_{12} & b_{13} \\ b_{12} & b_{22} & b_{23} \\ b_{13} & b_{23} & b_{33} \end{vmatrix}, \quad \dots, \quad
\begin{vmatrix} b_{11} & \dots & b_{1n} \\ \cdot & \cdot & \cdot \\ b_{1n} & \dots & b_{nn} \end{vmatrix}
$$

all be positive.

When $n = 3$, for example, the quadratic form can be written as

$$V = b_{11}x_1^2 + b_{22}x_2^2 + b_{33}x_3^2 + 2b_{12}x_1x_2 + 2b_{13}x_1x_3 + 2b_{23}x_2x_3$$

and Sylvester's condition becomes

$$
b_{11} > 0, \quad
\begin{vmatrix} b_{11} & b_{12} \\ b_{12} & b_{22} \end{vmatrix} > 0, \quad
\begin{vmatrix} b_{11} & b_{12} & b_{13} \\ b_{12} & b_{22} & b_{23} \\ b_{13} & b_{23} & b_{33} \end{vmatrix} > 0.
$$

Exercises

1. Use the theory of the "special case" to determine whether or not the origin is asymptotically stable for the given systems from Section 5.

   (a) Exercise 1;
   (b) Exercise 2;
   (c) Exercise 4;
   (d) Exercise 6;
   (e) Exercise 7;
   (f) Exercise 8.

2.* Use the theory of the "special case" to show that the origin is an asymptotically stable equilibrium point of the system

$$\dot{x}_1 = -3x_1 + 3x_2 - x_3,$$
$$\dot{x}_2 = -2x_1 + x_2,$$
$$\dot{x}_3 = -x_1 + x_2 - x_3.$$

3.* Use the Liapunov function that was found for Exercise 2 to determine the stability or instability of the equilibrium point at the origin for the system

$$\dot{x}_1 = -3x_1 + 3x_2 - x_3,$$
$$\dot{x}_2 = -2x_1 + x_2 + x_3^2,$$
$$\dot{x}_3 = -x_1 + x_2 - x_3 - x_1x_2.$$

7 The theorems of LaSalle and of Četaev

The theorems of Liapunov in Section 5, while basic, are not ordinarily the most useful tools for studying the stability of an isolated equilibrium point. In this section, we shall state, without proof, two basic and very useful results—the theorems of LaSalle and of Četaev.

Theorem 7.4. (LASALLE)† *In the system*

$$(7.1) \qquad\qquad \frac{dx}{dt} = X(x),$$

let the functions $X(x)$ be of class C' in an open region H containing the origin and suppose that the origin is an isolated equilibrium point of this system.

If there exists a function $V(x)$ of class C' that is positive definite in H and has the nesting property there, and if $\dot{V} \leq 0$ along trajectories, with the origin (that is, the null trajectory) the only solution of (7.1) for which $\dot{V} = 0$, then the origin is asymptotically stable and every solution in H tends to the origin as $t \to +\infty$.

---

† See LaSalle and Lefschetz, p. 58.

Corollary to LaSalle's theorem. If the conditions that $V(x)$ be positive definite and $\dot{V}(x) \leq 0$ be changed to read $V(x)$ negative definite and $\dot{V}(x) \geq 0$, the conclusion of the theorem remains valid.

When the conditions of this corollary are met, it is usually simpler to replace the original $V(x)$ by $-V(x)$.

To illustrate the theorem consider the system

$$\frac{dx}{dt} = y,$$

(7.2)

$$\frac{dy}{dt} = -2x - 3y - 2x^3$$

that has an isolated equilibrium point at the origin. We choose

(7.3) $V(x, y) = \tfrac{1}{2}(y^2 + 2x^2 + x^4),$

noting that V is positive definite neighboring the origin and that the origin is the only critical point of $V(x, y)$. Calculating $\dot{V}$ we have

$$\dot{V} = y(-2x - 3y - 2x^3) + (2x + 2x^3)y$$
$$= -3y^2.$$

Clearly, $\dot{V} \leq 0$, and since $y = 0$ is not a trajectory of (7.2), $\dot{V} \neq 0$ along all trajectories except the null trajectory at the origin. LaSalle's theorem obtains, and we conclude that the equilibrium point at the origin of the system (7.2) is asymptotically stable.

The student may wonder how the function $V(x, y)$ given by (7.3) was determined. This was done in the following way. To study the stability or instability of solutions of the second-order differential equation

$$\ddot{x} + p(x, \dot{x}) = 0,$$

where $p(x, \dot{x})$ is of class C' near $(0, 0)$ and $(0, 0)$ is an isolated zero of $p(x, \dot{x})$, it is helpful to write the equivalent first-order system

(7.4)
$$\dot{x} = y,$$
$$\dot{y} = -p(x, y).$$

It has been shown† that, for a wide class of functions $p(x, y)$, the function

(7.5) $2V(x, y) = y^2 + 2\displaystyle\int_0^x p(x, 0)\, dx$

---

† "On the Construction of Liapunov Functions for Certain Autonomous Nonlinear Differential Equations," *Contributions to Diff. Eqns.*, Vol. II, No. 1–4 (1963), pp. 367–383.

enables one to determine the stability or instability of the equilibrium point at the origin of the system (7.4). An application of (7.5) to (7.2) yields (7.3) at once.

Theorem 7.2. (ČETAEV)† *In the system*

$$\frac{dx}{dt} = X(x),$$

let the functions $X(x)$ be of class C' in an (open) sphere

$$H: x_1^2 + x_2^2 + \cdots + x_n^2 < k^2 \qquad (k\ constant),$$

and suppose that the origin is an isolated equilibrium point of this system. Let H_1 be a spherical region

$$H_1: x_1^2 + x_2^2 + \cdots + x_n^2 \le k_1^2 \qquad (k_1^2 < k^2).$$

If there exists a function $V(x)$ of class C' in H such that

(a) $V(0) = 0$;
(b) $\dot{V} \ge 0$ *along a trajectory of the system*;
(c) *the set of points for which $\dot{V} = 0$ contains no nonnull trajectory of the system*;
(d) *in every neighborhood of the origin there is a point x_0 such that $V(x_0) > 0$,*

then the origin is unstable, and a trajectory that is at x_0 when $t = t_0$ leaves H, as t increases.

Example.   Consider the isolated equilibrium point at the origin of the system

(7.6)
$$\dot{x} = y,$$
$$\dot{y} = x - y,$$

and the function

$$V = x^2 - y^2.$$

The derivative

$$\dot{V} = y^2$$

is positive except along the x-axis, where it is zero. The only trajectory for which $\dot{V} = 0$ is the origin. It follows that Četaev's theorem applies, and the origin is unstable.

---

† For a proof of this theorem, see N. N. Krasovskii, *Stability of motion*, Stanford University Press, Stanford, California (1963), p. 69.

Corollary to Četaev's theorem.   *If the conditions* $V(x) > 0$, $\dot{V} \geq 0$ *in the theorem are replaced by* $V(x) < 0$, $\dot{V}(x) \leq 0$, *the conclusion of the theorem remains valid.*

A generalization of formula (7.5) is available.† If the given system has the form

$$\dot{x} = f(x, y),$$
$$\dot{y} = g(x, y),$$

and if the equation

$$f(x, y) = 0$$

has the solution

$$y = h(x),$$

a useful function $V(x, y)$ to try is

$$(7.7) \qquad V(x, y) = \int_{h(x)}^{y} f(x, y)\, dy - \int_{0}^{x} g[x, h(x)]\, dx.$$

Example.   Consider the system (cf. Exercise 7, Section 5)

$$\dot{x} = -x + y,$$
$$\dot{y} = -2x - 4y.$$

To apply (7.7) we note that $h(x) = x$, and we have

$$(7.7)' \qquad V(x, y) = \int_{x}^{y} (-x + y)\, dy - \int_{0}^{x} (-6x)\, dx,$$

$$2V = 7x^2 - 2xy + y^2.$$

V is positive definite and

$$\dot{V} = -5(x - y)^2.$$

LaSalle's theorem applies, and the origin is asymptotically stable.

As a second example, consider the system (cf. Exercise 2, Section 5)

$$(7.8) \qquad \begin{aligned} \dot{x} &= x + y, \\ \dot{y} &= 3x - y. \end{aligned}$$

Here, $h(x) = -x$, and

$$V(x, y) = \int_{-x}^{y} (x + y)\, dy - \int_{0}^{x} 4x\, dx,$$

$$2V = -3x^2 + 2xy + y^2.$$

---

† L. R. Anderson and W. Leighton, "Liapunov Functions for Autonomous Systems of Second Order, *J. Math. Analysis & Appl.*, Vol. 23 (1968), pp. 645–64.

Along trajectories,

(7.9) $\dot{V} \equiv 0.$

From this result we conclude first that, by Četaev's theorem, the origin is unstable. We also note that (7.9) implies that we have accidentally discovered the trajectories of (7.8). They are the curves

$$-3x^2 + 2xy + y^2 = k \qquad (k \text{ constant}).$$

As a third example in the use of (7.7) consider the system

(7.10)
$$\dot{x} = 3x + 2y,$$
$$\dot{y} = -8x - 3y.$$

Here, $h(x) = -3x/2$, and

$$V(x, y) = \int_{-\frac{3}{2}x}^{y} (3x + 2y)\, dy + \int_{0}^{x} \frac{7x}{2}\, dx$$
$$= 4x^2 + 3xy + y^2,$$

a positive definite function. We have

(7.11) $\dot{V} \equiv 0.$

By Liapunov's theorem, the origin in (7.10) is stable. Further, (7.11) yields the fact that trajectories of (7.10) are the ellipses

$$4x^2 + 3xy + y^2 = k^2 \qquad (k \text{ constant})$$

and, hence, the origin is a center.

A nonlinear example is the system

(7.12)
$$\dot{x} = x + y,$$
$$\dot{y} = -2x - 2y - 2x^3.$$

The equations of variation are

$$\dot{x} = x + y,$$
$$\dot{y} = -2x - 2y,$$

and the origin is not an isolated equilibrium point of this linear system. The nonlinear term in (7.12) will then be the decisive factor in the stability or instability of the origin for the system (7.12).

Formula (7.7) yields

$$2V = x^2 + 2xy + y^2 + x^4$$
$$= (x + y)^2 + x^4.$$

We note that V is positive definite. Then

$$\dot{V} = -(x + y)^2,$$

and, by LaSalle's theorem, the origin is asymptotically stable.

Next, consider the system

(7.13)
$$\dot{x} = x + y,$$
$$\dot{y} = -2x - 2y + 2x^3.$$

Formula (7.7) yields

$$2V = (x + y)^2 - x^2,$$

and

$$\dot{V} = -(x + y)^2.$$

The corollary to Četaev's theorem applies [or take $2V = x^2 - (x + y)^2$], and the origin is unstable.

Finally, consider the system

(7.14)
$$\dot{x} = y,$$
$$\dot{y} = -x + x^2 y.$$

The corresponding equations of variation are

$$\dot{x} = y,$$
$$\dot{y} = -x,$$

for which the origin is a (stable) center, the trajectories being the circles

$$x^2 + y^2 = k^2.$$

Formula (7.7), however, yields

$$2V = x^2 + y^2,$$
$$\dot{V} = x^2 y^2,$$

and, by Četaev's theorem, the origin in (7.14) is actually unstable.

The classical concept of a Liapunov function has been largely modified to mean any function $V(x)$ that, together with its derivative $\dot{V}$ along trajectories, can be used to determine the stability or instability of an equilibrium point of a differential system. All such usages are now generally regarded as part of *Liapunov's direct method*. The chief difficulty in this method lies in the determination of an appropriate Liapunov function $V(x)$, particularly for $n \geq 3$.

Although many of the illustrations of and exercises in Liapunov theory in this book involve linear systems for simplicity, it should be emphasized that the most important application of Liapunov's direct method lies in the study of nonlinear systems.

Exercises

1. Use formula (7.7) to determine the stability or instability of the equilibrium point at the origin of the following systems:

    (a) $\dot{x} = 2x + y,$
    $\dot{y} = -5x - 4y;$

    (b) $\dot{x} = x - 3y,$
    $\dot{y} = 4x - 6y;$

    (c) $\dot{x} = 3x - 5y,$
    $\dot{y} = 2x + y;$

    (d) $\dot{x} = 3x - 13y,$
    $\dot{y} = x - 3y;$

    (e) $\dot{x} = -2x + y,$
    $\dot{y} = 2x - 3y;$

    (f) $\dot{x} = y,$
    $\dot{y} = 2x + y;$

    (g) $\dot{x} = 2x + 2y,$
    $\dot{y} = x + 3y;$

    (h) $\dot{x} = 2x + 3y,$
    $\dot{y} = 3x + 2y.$

2. Use formula (7.7) to determine the stability or instability of the equilibrium point at the origin of the following nonlinear systems:

    (a) $\dot{x} = y,$
    $\dot{y} = -2x - 3y + 3x^2;$

    (b) $\dot{x} = 2x + y,$
    $\dot{y} = -6x - 3y - 2x^3;$

    (c) $\dot{x} = 2x + 2y,$
    $\dot{y} = -x - y - 2x^3;$

    (d) $\dot{x} = y,$
    $\dot{y} = -x + x^2.$

3. Assume $b > 0$ and apply formula (7.7) to the system

    $$\dot{x} = ax + by,$$
    $$\dot{y} = cx + dy,$$

    where a, b, c, and d are constants. Thus, show that

    $$2V = \frac{1}{b}[(ax + by)^2 + (ad - bc)x^2],$$

    $$\dot{V} = \frac{a + d}{b}(ax + by)^2.$$

    Note that if $ad - bc > 0$, V is positive definite, and if $a + d \neq 0$, either LaSalle's theorem or Četaev's theorem applies. If $ad - bc < 0$, and $a + d \neq 0$, show that Četaev's theorem or its corollary applies. If $a + d = 0$, what are the trajectories?

4. The substitution $\dot{x} = y$ in van der Pol's equation

$$\ddot{x} + \mu(x^2 - 1)\dot{x} + x = 0$$

leads to the nonlinear system

$$\frac{dx}{dt} = y,$$

$$\frac{dy}{dt} = -\mu(x^2 - 1)y - x.$$

Using $V = x^2 + y^2$ show that the equilibrium point at the origin is asymptotically stable when $\mu < 0$.

5. Discuss the stability of the equilibrium point at the origin of the nonlinear system

$$\frac{dx}{dt} = y + x^3 + xy^2,$$

$$\frac{dy}{dt} = -x + x^2 y + y^3.$$

Take $V = x^2 + y^2$.

6. Discuss the stability of the equilibrium point at the origin of the system

$$\dot{x} = y,$$
$$\dot{y} = -4x - y - 4x^2.$$

7. Find a function $V(x, y)$ that will establish that the origin is unstable in the system

$$\dot{x} = y,$$
$$\dot{y} = 4x - y.$$

8. Show that the origin is an asymptotically stable equilibrium point of the system

$$\dot{x} = -2x + y,$$
$$\dot{y} = -2x + y - x^3.$$

9. Show that the origin is a stable equilibrium point of the system

$$\dot{x} = -x + y,$$
$$\dot{y} = -2x + y - x^3,$$

10. Given the system

$$\dot{x} = y,$$
$$\dot{y} = -6x - y - 3x^2,$$

use $V(x, y)$ determined by (7.5) [or (7.7)] and LaSalle's theorem to show that the origin is asymptotically stable.

11.* Consider the system

$$\dot{x} = y,$$
$$\dot{y} = z,$$
$$\dot{z} = -2y - 3z - 2x^3,$$

which has the origin as an isolated equilibrium point. Compute the characteristic roots of the associated equations of variation. Regroup terms to show that near the origin the function

$$2V = 3x^4 + 11y^2 + z^2 + 4x^3y + 6yz$$

is positive definite and that along trajectories

$$\dot{V} = -6y^2(1 - x^2).$$

Finally, use LaSalle's theorem to establish the asymptotic stability of the origin in a sufficiently small neighborhood of the origin.

12.* Consider the system

$$\dot{x} = -x + y - z,$$
$$\dot{y} = y - 2z,$$
$$\dot{z} = -x + 3y - 3z,$$

and show that the origin is an asymptotically stable equilibrium point using the Liapunov function

$$2V = 7x^2 + 18y^2 + 21z^2 - 4xy + 6xz - 32yz.$$

Answers

1.   (a) unstable;   (b) asymptotically stable;   (c) unstable;
     (d) stable;   (e) asymptotically stable;   (f) unstable;
     (g) unstable;   (h) unstable.

2.   (a) asymptotically stable;   (b) asymptotically stable;
     (c) unstable;   (d) stable.

3.   When $a + d = 0$, the trajectories are $(ax + by)^2 + (ad - bc)x^2 = k^2$.

5.   Unstable.

6.   Asymptotically stable.

11. LaSalle's theorem applies between the planes $x = -1$ and $x = 1$.

Supplementary reading

Supplementary reading

Birkhoff, Garrett, and Gian-Carlo Rota, *Ordinary Differential Equations*, 2nd ed., Blaisdell, Waltham, Mass. (1969).

Cesari, L., *Asymptotic Behavior and Stability Problems in Ordinary Differential Equations*, 2nd ed., Academic Press, New York (1963).

Coddington, Earl A., and N. Levinson, *Theory of Ordinary Differential Equations*, McGraw-Hill, New York (1955).

Courant, Richard, and D. Hilbert, *Methods of Mathematical Physics*, Vol. 1 (Interscience), Wiley, New York (1953).

Finizio, N., and Ladas, G., *Ordinary Differential Equations with Modern Applications*, Wadsworth Publishing Company, Belmont, California, (1978).

Hartman, Philip, *Ordinary Differential Equations*, Wiley, New York (1964).

Hurewicz, Witold, *Lectures on Ordinary Differential Equations*, Massachusetts Institute of Technology, Cambridge, Mass. (1958).

Ince, E. L., *Ordinary Differential Equations*, 4th ed., Dover, New York (1953).

Kaplan, Wilfred, *Ordinary Differential Equations*, Addison-Wesley, Reading, Mass. (1958).

LaSalle, J. P., and S. Lefschetz, *Stability by Liapunov's Direct Method, with Applications*, Academic Press, New York (1961).

Sansone, G., and R. Conti, *Non-linear Differental Equations*, Pergamon, New York (1952).

Widder, David V., *Advanced Calculus*, 2nd ed., Prentice-Hall, Englewood Cliffs, N.J. (1961).

Index